I0027167

BEYOND THE BOOKCASE

The Botanical
Lore of the California Indians

By
John Bruno Romero

Published by BTB Publishing

ISBN: 978-1-63652-284-5

THE BOTANICAL LORE OF THE CALIFORNIA INDIANS

JOHN BRUNO ROMERO

TABLE OF CONTENTS

PREFACE

Of all the books written concerning the Indians of North America, I don't know of one which treats of the Indians' great knowledge of medicine, the vast store which was theirs of plants and herbs which possessed curative and healing qualities, many of them far superior, even today, to the medicine used by the white physician.

There is a reason. In some instances the white man did not get the correct information from his Indian brother due to the latter's inability to make himself understood—this was, of course, also true of the former. Again, some information given was intentionally wrong due to the ill-feeling the Indian had for the white man. And again, many of those healing plants were held in such veneration by the Indians, that to impart their virtues to a white man was an unpardonable crime, and the punishment meted out to the offender was of the severest form.

I am an Indian, proud of it and of my forefathers, whose bitterness toward the white man was only too well justified. But time changes all things and bitterness and hatred never made for understanding nor happiness.

In this spirit I wrote this book, in the spirit of doing good. And in this I have the help and permission of my dear uncle, Chief Pablo, of the Mahuna tribe of Indians of Southern California, who permitted me to describe certain plants whose curative properties have been kept a secret by the Indians for over one hundred years. This is the first time they are made known.

The Indian, living close to and with nature—the greatest teacher of all for those who have eyes to see—became nature's most intelligent pupil. Gifted with the keenest observation **viii** and the ability to reason, he searched the discovered plants which nature herself had provided for any ailment, sickness, or mishap which might befall him.

I am sending this book out into the world not for fame, but as a messenger of goodwill and peace. May it be received in this spirit and accomplish its mission.

THE STORY OF THE INDIANS OF THE PACIFIC SOUTHWEST

A gruesome, terrible year, the year of 1825! The fatal year when thousands of Indians of the Pacific Southwest were destroyed by that merciless, frightful scourge, smallpox. And the tradition of its ravages is kept alive even today among the descendants of the few who escaped death.

The tribal herb doctors at that time were wholly unprepared to combat this disease which wrought such fearful havoc. Sweeping along the entire Pacific Coast, it exacted a heavy toll of human lives, so heavy, in fact, that the number of Indians destroyed exceeded that of the American lives lost on the battlefields during World War I.

The epidemic ravaged not only the Pacific Coast. It even spread over the adjacent territories, carrying death everywhere it struck.

Not until the end of that terrible year did some of the Indian herb doctors begin to devote their attention to the disease. And then, calling upon all there was in their knowledge of medicinal plants, the chieftains, accompanied by their medicine men, held a council at which the matter of curing this destructive disease was brought under serious discussion.

That memorable meeting took place in the world-famous Palm Canyon, bordering the eastern slopes of San Jacinto Peak—better known to us by its true native name of *Tahquitz*—situated at the extreme eastern part of Riverside County.

After the adjournment of the meeting the chiefs and medicine men dispersed, returning to their respective tribes to resume their regular duties, each one with instructions to study ways and means which would effectively eradicate the scourge so greatly feared by all Indians.

Now, among those who had attended the meeting was Senior Chief Andres Lucero, of the Mahuna tribe, who was looked upon by all Indians as the possessor of the greatest knowledge of botanical medicine, and a master-teacher among his fellow tribesmen who rightly considered him as being without a peer in his field—thorough research and experimentation. In his experimental work he was more successful than any other, having had many years of experience. In addition, he possessed valuable medical knowledge which for centuries had been handed down from generation to generation, each recipient becoming a true master in the field of Indian medical science.

The men trained in the science of medicine were those who had a natural aptitude and inclination for the care of the sick—that is, men worthy of being selected to be taught anatomy and the various ailments of the human body.

Evidence of this is to be seen in paintings and hieroglyphics found in our Indian caves, which, in due time, it will be my duty to use in photographic illustration in order to correct many erroneous interpretations existing in collections and writings.

To return to the beginning of our narrative, Andres Chino Pablo, while deeply pondering one day on the problem of what to give his stricken people, bethought himself suddenly of how in previous years he had treated violent fevers.

One was the fever which accompanied pneumonia and which was, at times, fully as dangerous as any fever known to mankind. But it was easily overcome through the administration of compounded herb steam baths.

In the course of time "Black Measles" made, by mere accident, its first appearance in the Pacific Southwest, again causing widespread terror among the Indians. However, Chief Andres Lucero had already decided what to do to check the epidemic. Standing calm and cool one morning and facing the rising sun, he called his people to him, everyone, young and old. Like a shaft of granite he stood, straight and erect, his eyes scanning carefully each member of his tribe. At last, with tears in his eyes, showing how deeply moved he was at the woe of his people, he announced his decision. His words were few:

"My sons and daughters, as the Great Spirit arises in the East, he comes to help us and he has given me the medicine and the power to cure all those of you who are sick. You will now go to the big cave where you will receive proper care and treatment. All of you shall go, even those who are not suffering from this devil's disease. And you will all be treated alike so that your blood shall be purified to guard you against the disease."

In other words, to render them "immune," in our language of today.

Now, one of the many caves in the San Jacinto mountain range, one which formerly had been used for religious studies in Chino Canyon, was converted into a cave-hospital and thousands of Indians were treated therein.

Due to the supposition, at the time, that the measles, then an unknown malady among the Indians, was really another form of smallpox—the diagnosis made didn't differ materially from the latter—the conclusion arrived at was to treat the disease accordingly.

A tireless and stubborn fight was waged by Chief Andres Lucero. Day and night he labored indefatigably, not only attending to his people, but also instructing other chieftains what to do and what to use, thus letting them share in his success in healing the sick.

Later on, the disease appeared in Twenty Nine Palms, in Borrego Valley, Indio, Coachella, Yuma, Temecula, Mojave, Tehachapi, Soboba—in fact, in all the small Indian settlements throughout the desert and mountain. Chief Andres Lucero was extremely satisfied with the results of his labors which had laid the cornerstone of knowledge and preparedness for the year of 1859, when a true epidemic of smallpox made its appearance again.

Investigations revealed that the scourge had started (*Temamaka*) to the north and had come (*Kichamba*) southward. Again the eye beheld the sad scenes of dead human bodies strewn along the valleys and mountains. The worst suffering was among the Indian tribes dwelling to the northward, who had shown a complete disregard of Chief Andres Chino Pablo, and paid no attention whatever to his messages and warnings.

However, the Chief stuck to his post until he saw his people safe. Those who died were comparatively few. The dead were those who, becoming panic-stricken through memories brought to their minds of former happenings, had fled into the desert. Had they but conquered their fears and obeyed the Chief's orders, nothing would have happened to them. Those were the victims of the plague—the deserters from the tribes. The smallpox was kept well under control in the settlements with only a two per cent death rate, which means practically nothing when compared with a previous death rate of fifty per cent caused by the same kind of plague. In some localities it rose to almost ninety per cent. Truly those were dark days indeed for the American Indians.

I shall now give, roughly, the medicinal plants used then and again in the same manner in the year 1918 when the "Black Plague," commonly known as "Spanish Influenza" was raging throughout the American continent and Europe. It was a strange coincidence, indeed, that a great many of those who escaped death from the enemy's bullets on the battlefield perished from the "Black Plague." The Indians, however, again won their battle against this frightful disease by the same means which had been employed against measles and smallpox with no loss of life whatsoever.

What a worthy and successful experiment it had proved to be under the wise guidance of Chief Andres Lucero, of the Mahuna Indians of

Guana-pia-pa! He was truly a noble character whose labors, love for his people, and self-sacrifice saved the lives of thousands of Indians, and without whose loyalty and kindness to other tribesmen, the American Indian in the Pacific Southwest would be an extinct race of people today, with no one to tell what had actually taken place in the wild mountain regions and the desolate desert, which were neither inhabited nor trespassed upon by the white race.

The Indians compounded steam-bath herbs for the cure of "Black Measles" and smallpox from the following:

	Indian Name
Artemisia tridentata	Ulu-ca-hul-vall
Larrea mexicana	Ato-col
Piperacea	Chu-co-pot
Eriodictyon glutinosum californicum	Ta-que-bel
Adiantum capillus-Veneris	Tal-wal

Herb tea given to the patients was made from the following:

Ephedra	Tut-tut
Sambucus pubens	Haa-saat

Please bear in mind that the Indian name Tut-tut bestowed upon the plant *Ephedra* means something that pertains to the very sacred in Indian medicinal art. This sacred Indian name befitted the plant for the great medical value it possessed and for the important part it played in stamping out the horrible smallpox plague of 1857.

Thus came about the saving of Indian lives.

Help us save and preserve the wild plants for the benefit of mankind.

CHIEF WILLIAM PABLO, III
Medicine Man

Mahum and Guana-pia-pa
meaning:
White Water and Palm Springs area, California

BOOK OF HERBS

Stomach disorders, worms, dysentery, diarrhea, etc.

SISYRINCHIUM ANGUSTIFOLIUM
(*Ind. Man-ta-ca*)

Commonly known as Blue Star Flower Grass. Found widely distributed along the rich southern slopes of the lower coastal ranges. Effective in the treatment of functional affections of the stomach. In the form of tea, the entire plant was used to eradicate all kinds of stomach worms. Flowering season from March to April.

DIPSACUS GLUTINOSUS
(*Ind. Vaah-se-le-coo*)

Known also as Monkey Flower, a plant not indigenous to any particular soil, grows abundantly in California in the lower coastal ranges, and also in the upper regions. Leaves, flowers, and stems were taken in the form of tea, and effected a thorough cure in severe cases of diarrhea.

VITIS VULPINA CALIFORNICA
(*Ind. Esq-urana-quat*)

Also called Wild Grape. It occurs mostly along rich river bottoms and marshy soils generally. Usually associated with Wild Berry Vines.

RUBUS VILLOSUS
(*Ind. Pick-lam*)

The Wild Blackberry. The roots of either of the two genera, boiled into a tea and given the patient, will afford permanent relief in mild cases of diarrhea. The roots may be gathered and used at any time of the year.

CAPSELLA BURSA-PASTORIS
(*Ind. Pa-sil*)

Also called Shepherd's Purse. Probably one of the most common of all plants in Southern California soils, growing throughout the year on irrigated lands and on arid soils. Boiled into a tea it is a certain cure for even severe cases of dysentery. No more than two to three cups should be taken. This little plant is a blessing to mankind and should be made use of. It is the medicinal queen, and surpasses all others in cases of dysentery and diarrhea.

HEDEOMA PULEGIOIDES
(*Ind. Mo-cash*)

American Pennyroyal. Considered as the greatest nuisance by farmers. As a curative agent in severe cases of dysentery it ranks next to *Capsella bursa-pastoris*. It is general throughout California, and blooms from August to September.

ANTHEMIS NOBILIS
(*Ind. Sa-mat-pl-ol*)

American Field Camomile. A very common plant growing everywhere in California. It was used extensively for babies suffering from colic, and also to regulate unsettled stomachs.

Painful congestion of the stomach.

MALVA ROTUNDIFOLIA
(*Ind. Mal-val*)

American Common Mallow, compounded with blossoms of California Wild Rose, or the seed.

ROSA CALIFORNICA
(*Ind. O-chul*)

American California Wild Rose. This beautiful wild rose-bush inhabits the coastal ranges, and may be found far inland along open spaces in heavy woodland abounding in rather rich and mulchy soil.

Used in cases of stomach fevers, the ripe seed is given in the form of tea to relieve the stomach clogged with food as well as in so-called cases of painful congestion.

MONARDELLA VILLOSA
(*Ind. Tah-lis-wet*)

Skirting for its habitat the high mountain lands, it is very seldom found anywhere else. It is a native of California, and is used chiefly for the relief of ordinary stomach-ache. When in bloom it is very fragrant and blossoms from late May till June. (American Horsemint—sometimes called Pennyroyal)

Fevers and constipation of the stomach.

ERYTHAEA MUEHLENBERGII
(*Ind. Co-oniche-la-wa*)

Its habitat is confined to a few localities in San Diego County, the coastal regions in Santa Barbara, Orange County, and in San Bernardino along the southern border of the Mojave Desert.

It was used in the form of an infusion in cases of constipation caused by fever of the stomach.

Fevers.

JUNIPERUS CALIFORNICA
(*Ind. Gla-wat-pool*)

American Juniper Berry. Its habitat extends from our high mountain ranges in northern San Diego County to Monterey County. The berries have a short season, ripening in these regions from late in July until early in September. They were used for making tea or simply chewed in cases of La Grippe fevers. They may be gathered, dried, and stored for future use.

Lung fevers.

PAEONIA BROWNII
(*Ind. Quel-ta-bat*)

American Wild Peony. It inhabits shady canyons growing only on deep, rich, decomposed mulch. The blossoms are of deep red color, and the blooming season lasts from May until June. Its roots bear a strong resemblance to young sweet potatoes and were gathered to be used in the form of tea for complicated lung fevers. The tea has a decidedly bitter taste.

Stomach fevers.

VERBENA HASTATA
(*Ind. Muy-u-vees*)

American Wild Vervain. Inhabits the lower coastal ranges and pasture lands. Its blooming period is from late in May until July. This plant is remarkable for bearing three different colored blossoms—in some localities white, in others, pink, and then again, blue—all this owing to the mineral soil formation. The root is used for complicated stomach fevers.

Fevers.

MORAJAUM
(*Ind. Saa-al*)

Grows along rivers and lake borders. This plant, being of a semiaquatic nature, resembles some of our wild orchids and blooms but a short season. The entire plant is used in cases of fevers complicated with headaches.

Eruptive Fevers.

SAMBUCUS PUBENS
(*Ind. Haa-sat*)

American Elderberry. Indigenous to the coastal regions. The yellow blossoms were extensively used by all Indian tribes only in cases of measles.

MIRABILIS CALIFORNICA
(*Ind. See-wish-pe-yack*)

American Four O'clock. The root of this plant served the same purpose as American Elderberry. Its habitat is Santa Barbara County, Calif., and it is not found anywhere else in a wild state.

There are no records that it was used for other eruptive fevers such as smallpox, scarlet fever, etc. These were introduced into this country later on by white European adventurers and settlers, the first cases being recorded in 1825. These diseases were greatly dreaded by the Indians, and whenever any of them contracted this malady, they would invariably vacate the locality they were in, and move many miles away to virgin country.

And, as a warning of danger to fellow tribesmen and to keep them away from the abandoned camp, all the rock mortars and clay pots were turned upside down and partly buried in the ground.

Plant poisoning.

GRINDELIA CUNEIFOLIA
(*Ind. Mie-chowl*)

Grows in alkaline soils and its blooming season is from August to September. The plant was used a great deal in cases of itching skin eruption caused by poison oak, and is a cure for such disorders. (American Gum Plant)

RHUS DIVERSILOBA
(*Ind. E-yal*)

Botanical Serum. This is the poison oak itself, of which the roots, during the dormant period, are cut and properly dried. When taken in the form of tea in a quantity of not more than four ounces, it will render a person immune against any further poisoning. This is a bona fide Indian formula.[1] Blooming season from May to June.

[1]I ask my readers not to try this serum pending further laboratory experiments. I plan to subject myself to exhaustive tests under scientific observation and to publish the results.

Ulcers and diseases of the skin and feet.

ANTENNARIA MARGARITACEA
(*Ind. Te-bish-samat*)

American Cotton Weed. Its habitat is all along the Southern California hill slopes, and the beautiful pearly flowers are used for ulcers and sores of the feet which fail to respond to treatment by other medicaments. The blossoms must be boiled. The liquid obtained is used to bathe the feet, and all parts of the skin affected. A very effective cure. The blossoms are also ground into a powder and applied to the part affected.

Burns.

SOLIDAGO NEMORALIS
(*Ind. Pa-co-se-cheeh*)

American Western Goldenrod. Its habitat is the river and creek bottom lands. Quite common in California, this plant has great healing power, especially in cases of old raw burns that have failed to heal properly, as well as major rotten ulcers. The leaves of the plant may be boiled and the liquid used to bathe the affected parts; while the pulp, as a poultice, is to be placed upon the burns and ulcers to promote disinfection and to hasten a rapid growth of new healthy flesh.

Solidago nemoralis has one other great virtue of considerable value. Two to three baths in a strong solution, prepared by boiling, will cure the Seven-Year Itch and free the victim from that terrible malady.

Fistulas and running sores.

PENTSTEMON CORDIFOLIUS
(*Ind. Ting-gi-wit*)

American Wild Fuchsia. A native of the coastal ranges, northward from San Diego County to Monterey, Calif., this dark-green shrub is very attractive to the eye. It bears an array of deep-red blossoms, well-formed in clusters, at the very tip of long slender branches. It was used as a poultice and a wash for fistulas and deep, pus-running ulcers.

Eruptive scalp diseases.

EUPHORBIA
(*Ind. Te-mal-hepe*)

A native of California, it is quite common on our inland fields. It is used for minor skin eruptions and scalp diseases. Used as a wash only. It blooms from early May to July.

AMBROSIA ARTEMISIFOLIA
(*Ind. Watch-ish*)

American Common Ragweed. Grows in abundance in swamps and along waterways. There are two distinct species of this worthy plant, the dwarf variety and the gigantic kind. Either may be used for the same purpose as mentioned for *Euphorbia*. In full bloom from July to September.

How to retain the natural color of the hair.

ARTEMISIA TRIDENTATA
(*Ind. Ulu-ca-hul-vaal*)

American Common Sage. Habitat, the California desert. For a good many years this plant has been used to restore the color of hair, but the method used and practiced is far from that of our people, the Indians. And this can be traced back to the misunderstanding of the people who first introduced this use of sage for that purpose. Of course, washing the hair with it as a scalp treatment will do no harm, because the plant itself has soothing and healing qualities; but to maintain the natural color of the hair, or to restore gray hair to its former color, one must drink the infusion when compounded in the way we know. The plant blooms from April to June.

Women's diseases.

ARTEMISIA CALIFORNICA
(*Ind. Hul-vaal*)

Habitat, the coastal regions. The infusion made from this plant was used a great deal in cases of vaginal trouble. Blooming period from March to May. (American Wormwood)

RAMONA POLYSTACHYA
(*Ind. Qua-seel*)

American White Salvia, playing a very important part in healing internally and removing particles of the afterbirth. The infusion from the roots was given to the patient to drink regularly in place of water.

LARREA MEXICANA
(*Ind. Ato-col*)

American Creosote Bush. Its habitat covers the entire length and breadth of the Mojave Desert, San Bernardino County, and Riverside. This plant was used in cases of cramps of the stomach due to delayed menstruation, and in cases of this nature not more than one half of a cup of the tea was drunk. This plant is in full foliage from May to October.

CHENOPODIUM AMBROSIOIDES
(*Ind. Epa-so-tee*)

Its habitat is in swamp bottom lands. The root of the plant was used in cases where the menstrual period had been overdue for five or as many as ten days. The plant itself has a rather offensive odor, but the boiled root is quite agreeable to the taste and very effective. The patient may drink as much of the tea as desired. Blooming season from March to late fall. (American Goosefoot)

CRACCA VIRGINIANA
(*Ind. Po-hiel*)

American Garden Rue. A common garden shrub introduced into this country at the beginning of the early mission days.

Although the odor of this plant is quite disagreeable to the sense of smell, the infusion is very rich in flavor and not bad at all.

Flesh-wounds, knife-cuts, etc.

ANEMOPSIS CALIFORNICA
(*Ind. Che-vnes*)

American Swamp Root. Habitat, swamps. This plant is plentiful in California—the territory where it grows wild could be measured in thousands of acres. When cut, dried, and powdered, it can be used for the disinfection of knife-cut wounds, and to draw and promote the growth of healthy flesh. (Spanish *Yerba Mansa*)

GRINDELIA SQUARROSA
(*Ind. Tanga-wet*)

Habitat, low, sandy loam soils. For above-mentioned purposes this plant is very valuable from a medicinal standpoint, as it makes all wounds respond quickly to healing, when used as a wash and for disinfection of cuts. A wet pulpy poultice must be applied to the wounds for quick results. The plant blooms from June to August. (American Gum Plant)

FRASERA
(*Ind. So-cat-llami*)

American Deer Ears. Habitat, the high sierras and coastal ranges. The infusion is used for the treatment of infected sores.

CARDUACEA
(*Ind. San-ca*)

American Green Sage. Its habitat is the Mojave Desert, San Bernardino County and north of the southern borders of the San Joaquin Valley. This plant is very valuable, being very powerful and of great medicinal use, and much attention should be given it by men of science.

The Indians used it universally in cases of serious and major wounds—the infusion being given the patient if symptoms of blood poisoning were present. Tetanus, commonly known as lockjaw, was easily overcome, thus eliminating the surgical operations so frequently resorted to by the medical profession. The infusion was also administered in cases of childbirth as a preventative of blood poisoning and gangrene with *Ramona polystachya*.

OPUNTIA
(*Ind. Tu-nah*)

American Cactus Pear. Its habitat, the desert and dry lands. This plant was fully as important as *Piperacea*. The large leaves were scraped of their thorns and a plug made out of the leaf, according to the nature of the wound, and inserted into it, healing it quite rapidly—a first-class piece of botanical surgery.

Healing.

PLANTAGO MAJOR
(*Ind. Pal-qua-ah*)

American Plantain. Its habitat is swamps and localities where there is abundant moisture. The plant, like many others, was used to dislodge and draw deeply embedded poisonous thorns and splinters from the flesh. The operation was quite simple. It consisted of applying a light coating of suet on one of the leaves, this was covered with another leaf and then placed, tied down firmly, over the thorn or splinter to be removed. It usually requires about 10 hours for the thorn to appear at the surface of the skin. The same procedure can also be used by persons who have accidentally stepped on a rusty nail—thus avoiding danger of blood poisoning. The simple poultice described above will prevent that.

CLEMATIS LIGUSTICIFOLIA
(*Ind. Chee-va-tow*)

The California Clematis is a sister plant of the Eastern Clematis, a very good healer in the treatment of skin eruptions, the infusion to be used as a wash.

Inhabits the mid-coast and inland ranges, and, to the east, the territory where Daniel Boone's activities played their part and took their place in American history. Nothing, however, is mentioned about this plant at the time the Indians were pursuing him in the wilds of Kentucky, and yet it was one of the strong vines of *Clematis* which enabled Daniel Boone to escape and save his life by cutting it with his hunting knife above ground and hurling himself far out, thus putting the Indians off his track.

Myself an Indian, I have always admired Daniel Boone for his cool presence of mind. He was brave and fearless, although not a showman like Buffalo Bill and others whose exploits were chiefly founded on personal motives.

Coughs, colds and sore throat.

RUMEX HYMENOCALLIS
(*Ind. Ca-na-ma*)

American Wild Rhubarb. Thrives in dead, sandy soils, and is very common throughout Southern California. The roots are long and bear a close resemblance to sweet potatoes. The infusion made from it has an acrid taste, and, when used as a gargle several times in cases of cough and sore throat, it will be found to give complete relief. The plant blooms in June and July.

PRUNUS SEROTINA
(*Ind. Is-lay*)

American California Wild Cherry. At home in the high mountain ranges. An infusion of the bark in spring or summer while the sap is running, or of the roots in winter when the tree is dormant, may be used for common coughs.

PRUNUS ILICIFOLIA
(*Ind. Is-lay*)

Holly-Leaf Cherry. Used for the same purposes as *Prunus serotina*.

SPIRAEA SALICIFOLIA
(*Ind. Ha-ba-ba-neek*)

American Queen of the Meadows. Its habitat is the low coastal ranges. The root of the plant was used for common coughs and chest colds.

EUPATORIUM PURPUREUM
(*Ind. Sa-ca-pe-yote*)

American Joe-Pye Weed. It was used for the same purpose as *Spiraea salicifolia* in localities where that plant couldn't be obtained, although the latter was greatly preferred for the extra medicinal qualities it possessed as a mild laxative. The root, when made into an infusion, is extremely pungent and rich in flavor, but agreeable in taste to most people.

MARRUBIUM VULGARE
(*Ind. O-o-hul*)

American Horehound. Its habitat is the woodland. Although the infusion made from the leaves and flowers is rather bitter, it is very good for ordinary coughs and sore throats.

Old dry coughs.

AUDIBERTIAS STACHYOIDES
(*Ind. Seel*)

American Black Sage. This plant is one of the most valuable of all for the cure of deep dry coughs of long standing, which have settled in the bronchial tubes. This does not mean coughs of two or three weeks' duration, but those which have existed for a period of from four to six months and which have, therefore, reached a chronic, dangerous stage.

The infusion was made full strength and given to the patient in small doses, hot—never cold—in the daytime, and one extra big dose before retiring.

Blood hemorrhages of the lungs.

DENNSTAEDTIA PUNCTILOBULA
(*Ind. Ma-ciel*)

American Hay-Scented Wild Fern. Its habitat lies in the high California mountain ranges. We are now coming to the tuberculosis line. Hemorrhages of the lungs, and common diseases which prevail to a great extent among people who, through neglect and irregular habits, intensify coughs and colds.

It was nothing to an Indian to overcome these maladies of the lungs, which in his case were usually due to accidental injury. This wild fern bears oil nodules on the crown of the root system and they are available only at a certain period, from May to June.

Coughs and asthma.

ERIODICTYON GLUTINOSUM CALIFORNICUM
(*Ind. Tan-que-bel*)

Commonly known as *Yerba Santa*, this plant proved to be possessed of great medicinal merits, and was very soon adopted by the mission friars for its outstanding qualities in the cure of coughs, asthma, rheumatism and pneumonia, being rightly considered as far superior in this respect to any of the other medicines brought by them from Europe. In fact, so great was the medicinal usefulness of these plants and hundreds of others known to the Indians, that they soon became the objects of study and investigation, which, however, met with failure, due to the severe punishment meted out to any and all Indians for divulging any secrets pertaining to the medical history of plants used by the tribes. A penalty which was sufficient to deter them from any further misdeeds in that direction, and which they always remembered. Quite a contrast to the modern, elastic laws of our present civilization.

ERIODICTYON CALIFORNICUM
(*Ind. Que-bel*)

American White Woolly Holly Plant is the sister plant of *E. glutinosum*.

Cathartics.

ERIOGONUM ELATUM
(*Ind. Pa-va-coneel*)

American Bottle-Weed. Its habitat lies in the volcanic regions of the Mojave Desert. This plant is rather peculiar in its growth, thriving on poisonous volcanic soils, where no other form of plant life can exist. The Indians of the desert regions used the plant as a physic, and it outranks *Rhamnus californica* in this respect. The mission friars overlooked this plant for the reason that none of them ventured that far into the desert, valuing their lives above everything else.

The infusion obtained from the plant was used in very minimum doses, and when unable to do that, a small branch was cut and a very small piece was chewed by the constipated person.

RHAMNUS CALIFORNICA
(*Ind. Hoon-wet-que-wa*)

American Coffee Berry. Its habitat is the canyons of high mountain ranges along waterway banks.

The bark was stripped off the trees, shade-dried and then ground in a *ca-wish-pat-os-vaal*, meaning the stone mortar and pestle generally used in those days, and even by druggists today, though made of different material. The prepared powder was used to a great extent at full strength in cases of constipation, and was administered in well-measured doses, but not in excess.

Owing to its medicinal properties this tree-plant was introduced into European countries where it gradually became the outstanding cathartic of all.

And this is the *Rhamnus californica*, the medicine of the Indians, named by Junipero De Serra *Cascara sagrada*—"Sacred Bark."

Kidneys.

EQUISETUM HYEMALE
(*Ind. Po-po-ot*)

Its habitat is confined to swampy lands. This plant is very fond of water, and attains a very vigorous growth under these conditions. Due to its aquatic nature, the plant, when fully matured, was gathered, shade-dried and an infusion made which was used solely in the treatment of prostate gland trouble. (American Horsetail)

EPHEDRA
(*Ind. Tut-tut*)

Its habitat is the desert lands of California, northwestern Arizona, and Nevada.

This evergreen, shrubby plant was held in high esteem by all the

Indians, and a good supply of it was always kept on hand for general use. The infusion made from it was used regularly to flush the kidneys. The tea is of a very delicious taste. A person cannot help liking it, and it also helps to purify the blood. (American Tea of the Indian)

PELLAEA ATROPURPUREA
(*Ind. Cala-wala*)

American Purple Cliff Brake Fern. Its habitat: the high mountain ranges. This useful little fern grows abundantly on most of the limestone formations and is seldom found anywhere else. Like *Ephedra* it makes a delicious tea, which is used more or less for the same purpose, to flush the kidneys and to tone and thin the blood in severely hot summer weather as a preventative against sunstroke.

Blood pressure, sunstroke.

ERIOGONUM ELONGATUM
(*Ind. Te-ve-na-wa*)

This plant is an inhabitant of the Mojave Desert. There are two different varieties, one of them being quite common on arid lands and side hills along our coastal highways. The other is the best, however, and, as a blood tonic, compares very favorably in medicinal worth with all others recommended.

The latter was used by the Indian for special cases of high blood pressure and hardening of the arteries.

It was generally used by Indian runners, and taken before and after a long-distance run over rough country.

Sedatives.

VERONICA OFFICINALIS
(*Ind. Ca-wish-hubel*)

American Speedwell.

MENTHA SPICATA
(Ind. Ga-vish-ho-ba-jat)

American Garden Spearmint. Habitat, the lower marshy coastal regions.

ILYSANTHUS BRACHIATUS
(Ind. Samat-hubel)

American Mountain False Pennyroyal.

MENTHA CANADENSIS
(Ind. Samat)

Both *Ilysanthus* (above) and *Mentha Canadensis* inhabit the high mid-coastal ranges and are frequently found lining the borders of mountain streams in beautiful settings of wild ferns. (American Mint)

MICROMERIA DOUGLASII
(Ind. Ya-mish-hubel)

Mint Family. A rare plant and found only in a few localities on the mid-coastal ranges, as in Orange County, San Juan Capistrano, at Hot Springs, situated on the southern slopes of the Trabuco mountains, Los Angeles County, Fish Canyon, Pasadena in Santa Barbara County, the heavy woodlands of Montecito Valley and in the Old Spanish Grand Rancho, San Leandro. It is also found northward as far as San Francisco at Angel Islands. The infusion was taken to soothe the nervous system in cases of insomnia.

Catarrh of the head and nasal chambers.

PLATANUS OCCIDENTALIS
(Ind. Ci-vil)

American Sycamore. It is an inhabitant of the California mountain ranges. The underside of its leaves bears a very fine yellowish moss, which the beautiful little hummingbirds like to use for building their tiny nests.

In fact, they prefer it to any other material on account of its extreme softness.

These leaves are valuable as an effective cure for old chronic cases of catarrh, when the catarrh has passed into internal ulcers, which continually discharge material of an offensive odor.

The moss scraped from the underside of the leaves, carefully and patiently enough to have a sufficient supply to compound it with the dry powdered yolks of the eggs of quail and an infusion of *Andromeda polifolia* was also made and used as a nasal douche, to cleanse the conduits, followed afterwards by sniffing the powdered compound before retiring for the night.

HELENIUM AUTUMNALE—HELENIUM NUDIFLORUM (*Ind. Pe-bah*)

American Sneeze-Weed. Both inhabit swamps and mountain springs.

ANDROMEDA POLIFOLIA (*Ind. Ho-bef-zo-bal*)

American Moorwort. This is found only at very high mountain altitudes.

Toothache and pyorrhea.

ACHILLEA MILLEFOLIUM (*Ind. Pas-wat*)

American Yarrow. This plant bears a strong resemblance to the Wood Betony, which is poisonous, and both may, in their wild state, be found side by side in the same locality. It is indigenous to the mid-coastal range woodlands.

Persons suffering from a severe toothache can cut the tips of the leaves of *Achillea millefolium*, roll them into a small pellet and insert it into the cavity of the aching tooth. You will be surprised how quickly the pain disappears.

ERIOGONUM
(*Ind. Pas-vaat*)

This inhabitant of the Mojave Desert, with its golden yellow flowers, is a treat to the true lover of nature. Its blooming season is in its full glory during the month of October when everything else in the way of flowering trees and shrubs is lying dormant. Thousands of acres in the desert may be seen carpeted with this golden color, blending with that of autumnal foliage and geological soil formation which also glows in many tints, offering wonders of inspiration to the artist—the greatest interpreter of the works of nature next to the botanist, geologist and naturalist.

For centuries *Pas-vaat* has been used by the Indians to keep the roots of the teeth and the gums in a state of health. Whenever pyorrhea was present and the teeth threatened to become loose, an infusion of the flowers and leaves of the plant was made and regularly used as a mouthwash. Although its taste is very bitter, holding the liquid in the mouth for a few minutes daily will prevent and cure pyorrhea and tend to the firm setting of the teeth.

QUERCUS RUBRA
(*Ind. Qui-neel*)

American Red Oak. The juice obtained from the bark is a very efficient means for straightening and setting loose teeth, but it has no effect on pyorrhea.

PERSEA and VANILLA PLANIFOLIA

American Avocado, or Alligator Pear. Our Indian brothers of the North Central American states use the seeds of the avocado in the treatment of pyorrhea, although only in the form of powder. It is very good and efficient for toothache, as is also an infusion made of the vanilla bean.

Fever and chills.

ROSA GALLICA
(*Ind. Mal-va-pol*)

American *Malva rosa*. This rose tree has to some extent been the subject of discussion among some of our botanical explorers and the result was always one of indecision.

Now let us look back a few years before the founding of the California Missions, and thus settle the dispute for all time. Twelve miles eastward of the Santa Barbara Mission is a small village by the name of Carpenteria, and at one time, this village was one of the largest Indian settlements in existence.

Before the arrival of the mission friars this place was a dense forest of the giant elm, *Ulmus pubescens*, a tree which is very soft and easy to work and the Indian settlement became the scene of great boat-building activity. The biggest and best trees were selected, hewed out and shaped into boats, and their boats were later used to navigate from Santa Barbara across the channel to Santa Rosa Island and other points. The Indians traded with the inhabitants of these islands, and thus they attained a great deal of magnetic iron in exchange for the wild products of the Pacific coast mainland. The magnetic iron was of general use among the Indians, being made into hammers, axes, scraping and cutting knives, fighting weapons, etc., all made in true Indian mechanical design.

Other valuable rock materials were also traded for, such as obsidian for arrowheads and small mortars, metates made of the volcanic rock found on the islands, and also ironwood. These were all materials much preferred by the Indians to those of the mainland, which were rather unfit to shape into stone utensils because they did not have the proper cleavage. The Santa Barbara mountain ranges offered none of these materials—only the minerals, Ullmannite, Oligocene rock and red jasper which sometimes served as passably fair substitutes.

Referring again to the trade articles of the island of Santa Rosa, the Indians, like their white brothers, liked to change and use different medicines. While navigating the rough sea across the Santa Barbara channel in their little boats, some of them would sometimes catch cold from getting wet and being exposed to cold winds. When reaching the islands some would have a high fever and chills, and then aid was given them in the form of *Malva rosa*, this being the plant used to break up the fever. And it will do so if properly administered. The Indians had so much faith in its value that they brought some of the seeds to the mainland where they were planted.

Later, on its introduction among other Indian tribes of the coastal belt, the plant found its way north- and southward until the coming of the California mission founders. They learned of the plant's medicinal value through Indian information, and were only too glad to adopt it for their own use—that plant and many others which proved superior to those brought by them from Europe and which they then discarded.

Thereupon the friars adapted themselves to the care and use of the herbs of the Indians.

This is the story of the *Malva rosa* after which the island of Santa Barbara bears its name *Santa Rosa*, or "Holy Rose," and botanically *Rosa gallica*.

Fractures.

ULMUS PUBESCENS
(*Ind. He-wa-wa*)

American Elm. We have seen that the beautiful elm was used by choice as a light, soft boat-building material. It played also a very useful and important part in the adjustment and healing of broken and fractured arms or legs. The work was very simple and effective. The patient was placed in bed, or what was known in those days as the *un-wet*, meaning bearskin mattress, to lie down and rest till the Indian runners returned from the forest with the stripped bark of the elm, which was very carefully selected and had to be free of woody knots, with the inner side of the bark as smooth as silk. These large strips were cut to mould and fit clear around the broken bones, then tied with wet buckskin. This was done to allow contraction of the buckskin with that of the green juicy bark of the elm, while the fevered and swollen joint absorbed the juice of the bark.

Care was taken to add more juice extracted from the tree to the bark strips to prevent quick contraction which would be very painful, due to swelling and counterpressure from the drying bark. The time involved in healing broken bones could well be considered two thirds of the time taken under the hands of modern skilled surgeons.

In parts of the country where the elm wasn't available a freshly killed rabbit, its skin quickly removed and slipped onto the broken joint served equally as well, only it required more time to heal.

Blood specific, purifier and tonic.

FOUQUIERIA SPLENDENS
(*Ind. Gaiesh-pohl*)

American Desert Candlewood. Spanish, *Ocotillo*. This plant's habitat is the southeastern wings of the Mojave Desert, and the locality best suited to its growth is Borrego Valley at the northern border of San Diego County. This great valley, at one time very rich and fertile, was used by the Indian tribes of Chief Hobo-yak of Ca-we for the raising of considerable livestock. This particular spot commanded an extensive view of the desert territory, as well as the mountain peaks surrounding it. From the top, a clear view could be obtained toward north, northeast and southeast to the Mexican border, and it afforded a natural fortification for the aggressive Ca-we Indians.

There still remain a few of the sand-dune forts heavily overgrown with creosote bushes. These forts are deeply recessed, formed in the shape of a horseshoe, its outlet serving as an entrance at the same time. Pointing northward toward the high mountain ranges, the graves, or burialground of the Indians, are located just outside of the fort and a few feet to the left from the outlet on a well-arranged plot of ground.

The interment was simple. After a grave had been dug, it was filled with dry wood and set on fire until a good bed of charcoal was attained; after which, the defunct together with his belongings was placed upon it and left to burn as the grave was covered.

Cremation took place slowly but surely.

Little evidence was left for the grave robber or the anthropologist. And still less for the archaeologist, as the defunct's pottery and rock mortars also were disposed of by being broken into thousands of pieces, and then scattered over the grave.

There is more to be said about this historical valley in that it was the main artery of caravan travel of the intrepid American and Mexican pioneers, and of the Spanish explorers, who all in turn were met and held up by the Indians for information. Those not offering resistance were allowed to pass through the territory unmolested—provided, of course, they wouldn't hang around the valley.

However, quite a number of bloody battles took place there, between Indians and whites, as the latter, having had some experience with hostile Apaches when crossing the desert, misunderstood the signals of the Ca-we Indians. Instead of waiting to see what they wanted, the leading scouts of the caravan ordered every man to stand by and, as the Indians came near, they opened fire on them. This didn't prove of benefit to the pioneers and only resulted in disaster, thus putting an end to their journey.

Even to this day many of their belongings are kept well secured in the bosom of Borrego Valley. Sand dunes uncovered now and then by desert winds prove there is a possibility of their all being recovered in time and preserved as historical relics of days gone by.

At the northwest end of the valley are the historical camp grounds of De Anza, the Spanish explorer, who was the only one ever known to have put up camp in the valley in those days.

And this marked the beginning of darker days for the Indians. First came the Spaniards, then the Mexicans and last the American gringos. They all passed there, but none had the least suspicion of what was to take place in the future. De Anza, while resting up for a short period, explored among the surrounding mountains in search of a pass which would lead westward. In this the Indians helped him by pointing toward the Pacific coast. De Anza reported to his superiors the finding of emeralds, but in this he was mistaken as the precious stones found in those mountains are mostly tourmalines in a great variety of tints, including green. This probably accounts for De Anza's naming them emeralds. They also occur in red, pink, yellow, white and black, with spodumene crystals in violet and purple.

The floor of the valley also yields carnelian milky quartz, bloodstones, as well as fossilized wood and jasper. These beautiful precious stones have found a place among the gems of the world and are well-known among mineralogists. In the traditional history of the Indians, De Anza is still remembered for the discipline and control he kept over his men, a fact which greatly facilitated his progress. And his goodwill and kindness toward the Indians were of the purest humaneness, standing out brightly as attributes of genuine manhood. Very few Spaniards could boast of such qualities—they ordinarily were brutal, and their history was written in blood in those days.

The Mexican soldiers who arrived later, by way of this valley, committed frightful excesses among the Indians, even assaulting and outraging the fair daughters of Ca-we. A horrible tragedy marked the sequel to those days of terror and bloodshed.

Then, in the rising dust on the northeastern horizon appeared a newcomer, General Kearny, in command of his gallant dragoons, following the course De Anza had taken. He had no trouble making the pass as his men were under strict military discipline and the blue uniforms worn by officers and troopers caused much admiration among the Indians. Little did they know, however, that behind this army was a power and authority which could be exercised without cause or provocation in the name of the government of the United States—and that was exactly General Kearny's mission: to retake the lands held by the Mexicans as their own at San Pasqual. He became engaged in a heated battle with Mexican soldiers and civilians and his casualties were so heavy that he faced defeat. As a last resort, he appealed to the Indians chiefs for help, and the Ca-we Indians, remembering only too well the atrocities and brutalities committed upon their women by the soldiers of Mexico, granted the General's request and dispatched five hundred warriors to aid General Kearny.

Arriving at the battlefield, the Indian braves attacked the Mexicans from all sides, thus saving the day for General Kearny who had but a few of his men left.

As to the kind of report he made to the War Department, and whether he took all the credit for having won this bloody battle whereas, by rights, the Indians alone were responsible for the victory, is not known to us. But we have every reason to believe that the report was in his favor in order to shield the command from embarrassment were it to become known that Indian aid and tactics had been the deciding factor. At any rate, up to the present time, Indian records don't show any letter of acknowledgment ever having been received from the United States Government for the service rendered General Kearny.

Instead of receiving at least a "Thank you," they were from that time on looked upon as "easy," believing anything that was told them—and this was proved to their own satisfaction in 1851-52. No doubt some of our fellow citizens do not know what this means. Therefore, to elucidate a little, let me say that this was the time when a benevolent government sent a very able regiment of politicians out here in command of a few army engineers.

These engineers executed their commission in a first-rate manner and negotiated and signed 18 treaties with the signatures also of 400 Indian chiefs. And the Indians turned over more than one hundred million acres of fertile lands to the government, leaving a reserve of ten million acres to the Indians. The larger acreage was to be paid for, as per duly signed agreement, at the rate of $1.25 per acre, and to this day the Indian is still waiting for it. The politicians, however, lost no time but took over as much as they could of these rich Indian lands and eventually the entire open country was known as the Ranch of Senator "This," or Senator, Congressman, or Colonel "That."

That the longest road has a turning was proved by the battle of San Pasqual: it opened the Indians' eyes to many things—such as, that had they but stayed out of it or used our forces otherwise, a different story would be told. The Indians were expelled from the Borrego Valley. We need no explanation here that they were good strategists and fighters and their holdings there were needed for the purpose mentioned. The valley has been donated by these patriotic and liberal politicians to the State of California and, by law, has become its property, and a state park.

And from high on its mountain top, the men of Ca-we may look down into the valley and think of its past history with an ache in their hearts. Only a few make a yearly trip there, for the purpose of gathering a goodly supply of medicinal plants and the *Ocotillo* or Indian Poinsettia Candlewood.

Blood tonic.

SALIX WASHINGTONIA
(*Ind. Ke-cham-ka*)

American Willow Tree. This is a common inhabitant of most of our swamps and rivers and is occasionally met with far inland. Like the elm tree, it is very fond of water, and both species may be found growing together in the desert or dry mountain gorges. We should, therefore, always look for such places when in want of water—for it is sure to be close to the surface and can be had by digging a few feet underground.

The Indians, when traveling across country, always stopped at such places where these trees were growing, and water was obtainable very quickly in sufficient quantities for domestic use. Don't get discouraged if you don't find water seepages on the surface. You must remember that the elm and the willow growing around that ancient fountain consume most of the water coming upward to the surface.

These sources of liquid supply were once the watering places for migratory birds flying east or northward, and by means of their droppings these trees became habitants of these wild regions.

Concerning the medicinal value of the willow tree, the Indians would take the leaves and pound them in the stone mortar, after which the pulp was put into water where it was allowed to remain a few hours. The liquid was then taken like any other beverage, the patient first having taken a bath to prevent a cold, and to keep the blood at the proper temperature.

The tea has a decidedly bitter taste. During the dormant season of the trees when the leaves were not available, the bark was stripped from the tree and served the purpose just as well.

TABARDILLO
(*Ind. Pees-wel*)

This strawberry-leaved plant grew in our time in abundance throughout the lower coastal ranges and valleys; but when the white pioneers in the early days began clearing the land for agricultural purposes, it marked the destruction of this and other valuable plants also.

Therefore, this plant is now totally extinct in the lower coastal plains, but still available in a few remote places in the higher ranges, where it is gathered by the Indians for home use.

Tea is made from the plant, and all the members of the family partake of it for at least one full month each year.

The infusion made from the root is of a clear, wine-red color, and is extremely beneficial as a blood tonic.

HOSACKIA GLABRA
(*Ind. Su-cot*)

American Deer Grass. Inhabiting the high coastal ranges, this beautiful little flowering grass when in bloom, during the month of May, appears to be almost a massive bouquet of carnations among the greenery of the mountain slopes.

Deer are frequently found foraging in places where the *Hosackia glabra* grows, as they are very fond of the grass, and the Indians, as a rule, pick such localities to hunt in.

The infusion made from the plant is rather aromatic and, if taken regularly, it will help to build up the blood.

Antidote.

PIPERACEA
(*Ind. Chu-co-pat*)

Habitat, rich northern mountain slopes. It grows mostly in underbrush, but is sometimes found on cleared land. It was gathered, dried and then ground in the rock mortar to a very fine powder, for use when our people exchanged poisoned arrows for bullets on the field of battle. Our poisoned arrows were more effective than bullets, as a scratch would send an enemy to eternal rest. Today we still use this plant in the treatment of trachoma, and rattlesnake, black-widow and scorpion bites. (American Pepper Plant)

Poison.

ESCHSCHOLTZIA CALIFORNICA
(*Ind. Tee-snnat*)

American Golden Poppy. The true native flower of our western domain—the flower of the beautiful, rich golden color. However, being a poisonous plant, the poppy fields were of no use to the good morals and medical practice of the Indian doctors in the days gone by. But the notorious witch-doctor, having more of the devil in him than of anything else, made general use of this plant to compound some of his poisonous medicines in his irregular and evil practice. This may be strange to learn, but we had its counterpart among other races—the white race included—where witchcraft with its incantations, love potions and straight poisoning of humans played quite a role. And so it was here among the Indians. For instance, if a woman became enamoured of a certain man and the object of her affections paid no attention to her advances, whether his regards were already placed in some other quarter, or for other reasons, this was sufficient for the woman who, having had her immoral desires thwarted, resorted to the witch-doctor. He would compound a preparation with the poppy as one of its main ingredients. This was then given to the innocent victim with the help of a second or third party, either in a solid or liquid form, and in less than twenty-four hours after partaking of it the poor fellow's mentality became befogged and powerless, and was thus easily controlled by the evil woman.

When such a case was brought to the attention of the chief of the tribe, orders were given to bring the malefactors in to be tried before the rest of the chiefs.

The leaders of this evil practice, if found guilty, were condemned to exile, and, under a guard of warriors, taken far out into the desolate desert with the death penalty hanging over them, should they ever return.

The poisoned victim was placed in the hands of the tribal doctor who gave him an antidote which counteracted the poison given by the witch—something which no white doctor has been able to do in spite of his knowledge of medical science and chemistry. Cases of such nature have happened among Indians and yet the patient's normal state of mind and health was restored.

ASTRAGALUS MOLLISSIMUS
(*Ind. Po-gat*)

American Locoweed. A good many years ago this poisonous plant was powdered and used to dope race horses, as the Spaniards were very fond of this sport of kings. And since they were the great landowners when the Americans began to come in, this sport became more popular than ever. Money was wagered against large tracts of the Mexican-Spanish lands in behalf of their favorite horses, and at times amounted to many hundreds of acres of the richest and most fertile land. It was this very poisonous *Astragalus mollissimus* that was responsible for the transference of large Spanish and Mexican grants into the hands of Americans.

The Indians, being well-informed and cognizant of the fraud being perpetrated at the expense of their good friends, revealed to them the tactics employed and, for a small compensation, offered to recover for them some of their losses. This offer was gladly accepted and, in consequence, the Mexican landowner would again challenge the former American winner, whereupon a date was set for the race.

Under cover of night the Indian would watch over the horse that was to run until the day of the race, when he would appear in order to redeem his promise. Although this may cause surprise, it was only a little Indian trick, playing its part of revenge on deceiving, dishonest persons.

It was clean, honest revenge, not requiring the poisoning of one of our most highly valued domestic animals, and this is how it was done.

The Indian rider concealed under the bosom of his buckskin shirt two pieces of skin, one from a fresh bearskin and the other from that of a mountain lion. When both riders were lined up for the race and right at the moment of taking off, the Indian, with a quick jerk, would pull both skins—which were hanging on a string—from under his shirt and his opponent's horse, quickly scenting them, would stop and balk, throw up his head and look fearfully around in all directions.

The Indian made the wire easily and thus the other horse's deceiving owner lost the race and also his ill-gotten gains, proving again that crime, in any form, doesn't pay in the long run.

Hair tonic, hair and scalp diseases.

LARREA MEXICANA
(*Ind. Ato-col*)

American Creosote Bush. I have mentioned the use of this shrubby bush in the beginning for another disease. It was used by the Indians for various maladies which I shall describe toward the end of the book.

The plant was one of the principal herbs used to eradicate dandruff and the infusion, when used as a hair-wash once a week for a period of about two months, will be found enough to rid a person of dandruff thoroughly. Its one drawback, if any, is that it will make the hair coarse, although on the other hand, very strong. It was also used as a disinfectant and deodorizer of bodily odors by sponging the body with the full-strength infusion.

LOPHOPHORA WILLIAMSII
(*Ind. Am-mool*)

American Desert Agave or Button Root. Indigenous to the Mojave Desert, San Bernardino County, East Riverside County and the Borrego Valley situated at the northern end of San Diego County, and thence southeastward to the Mexican border. The Indians made special trips to places where this plant grew, and spent several weeks at a time in harvesting it. The root is very important and yields the proper ingredients to compound one of the finest hair tonics known—greatly superior to the best on the American market of today.

After gathering the plants, the Indians roasted the leaves in covered underground rock pits. After this was done they underwent a sun-drying process, following which they were cut into small square pieces and stored away for winter use in hermetically sealed earthen *ay-as*.

The plant was put to many other uses which I shall not divulge at this time, as it would probably invite total extermination of the desert agave, as occurred in the case of the bison and the messenger dove.

Ringworms and scalp germs of the hair roots.

MICRAPELIS MICRACARPA
(*Ind. Yal*)

American Thorny Cucumber. This beautiful vine is very common through the lower and upper coastal ranges. Wherever you may go you will see the thorny cucumber. It is very bitter to the taste and the seeds are very pretty, running into the various shades of white, yellow, gray, black, olive-green, brown, and red.

In the '90s the white people discovered that they could be worked into portieres, hanging ornaments, etc., as they looked so beautiful and the fad became so strong that young and old would search for these seeds all over the country with the result that the plant became almost extinct, the usual net result when the white man took a fancy to certain plants or animals. However, in this instance, kind Mother Nature hid a few seeds away among and under the carpet of autumn leaves which in time replaced the destruction wrought by the white man.

One should know that we, the so-called savages, never destroyed wantonly, and when we gathered the flowers, seeds, leaves and bark of plants, we did so for a useful purpose, not to strew them along the highways as the custom is among certain humans of today.

Micrapelis micracarpa was used in the treatment of ringworm in the epidermis. Juices extracted from the rind of the thorny cucumber were rubbed into afflicted parts—a sure and effective remedy.

For the treatment of diseased scalps and hair roots the oil extracted from the seeds is massaged into the scalp and thus prevents the falling out of the hair.

The juice of *Micrapelis micracarpa* will remove human bloodstains, one of the most difficult stains to remove. The Indian made quite frequent use of it—when returning from the battlefield or a hunting trip—to remove such stains from his buckskin clothes.

The soap of the Indian.

CUCURBITA FOETIDISSIMA
(*Ind. Meh-hish*)

American Wild Gourd. This plant, inhabiting the arid soils of the coastal plains, is very hardy and at its best in worthless ground unsuitable for agricultural purposes. Thanks to Mother Nature, by thriving in otherwise dead soils, it was assured of continuous preservation.

The Indians regarded it highly as being useful in what washing they had to do toward keeping their buckskin clothes and blankets clean. Very soon after the vandals from Spain invaded our country, they adopted this plant in place of their greasy soaps. When the gourds were fully matured they were gathered and put away in the shade to dry for winter use. For the day's washing, one gourd was put in about ten gallons of water to help bleach the clothes, while large, square pieces were cut from the roots (which grow to an immense size) and were used as root soap bars for rubbing the clothes laid on a smooth log or a flat rock. Another plant similarly used was the fibrous bulb of the

CHLOROGALUM POMERIDIANUM
(*Ind. Mo-cee*)

A member of the Lily family, known as Soap Plant.

SAPONARIA OFFICIANALIS
(*Ind. Yu-look-ut-hish*)

American Soapwort. Inhabits the woodlands of the lower coast regions. The juice extracted from the roots of this plant makes an excellent hair tonic and cleanser, giving the hair a beautiful, brilliant gloss and, as a shampoo, it exceeds most of the modern hair tonics which, in my opinion, are more or less injurious to the hair roots due to an excess of alcohol in the preparations. Blooming season: May to June.

Its leaves were also of much use in cases of pains of the spleen. They were ground into a pulp and applied as a poultice-plaster directly where the pain was, and kept there for from three to four hours.

BRODIAEA
(*Ind. Meh-wahot*)

Is an inhabitant of the lower coastal regions, and is one of the other plants serviceable for making an excellent shampoo for the hair, the bulb and flowers being used for the purpose. Blooms from March to May.

Protection against lightning.

CEANOTHUS DIVARICATUS and PINUS SABINIANA

(*Ind. O-Oot*) (*Ind. Wa-at*)

The California Lilac—Spanish *Chaparral*—an inhabitant of the mountain ranges, blooms like the common lilac and is very beautiful.

Attracting every nature lover, this gigantic shrub is possessed of a strange power, which up to the present has escaped attention, due to its having been kept secret among the Indians for centuries.

As I am writing Indian legends through botany, I, for the first time, shall reveal the virtues of this shrub and it should prove of the greatest interest to those engaged in scientific research, especially in geology and botanical chemistry. And if the shrub is used as the Indians used it—and still do—it will help to safeguard life when thunder and lightning are at their worst at timber line of high mountainsides and peaks.

These monumental cones were at one time, in the early Archean period, surcharged with volcanic activity, which was responsible for the enormous iron deposits stored in the bowels of the earth. Electrical storms, being attracted by the iron, burst at times there with terrific violence, causing a truly phenomenal disturbance; while lightning bolts strike the iron bodies, completely magnetizing them, turning them gradually into the mineral, magnetic iron.

Now, let us bear in mind that cone-bearing pines and the *Chaparral* grow side by side on and over these magnetic iron deposits. Of the two, the pine seldom escapes destruction whereas the *Chaparral* deflects lightning. At the harvest season of the pine nuts, when these storms were very frequent, precaution had to be taken against the danger of being struck by lightning, and the *Chaparral* was the tree chosen in the pine forest by the Indians as furnishing both safety and shelter.

Now, some of our Red fellow men, through ignorance and a belief in miracles, were superstitious and quite sincere in their belief in the efficiency possessed by the magnetic iron mineral, and in its inherent supernatural powers to guard their homes against prowlers and thieves. So, much of it was gathered after its discovery and placed inside the homes to protect them from such intruders. Therefore, whenever a man was known to be in the possession of this mineral he became the object of much reverence and respect by wrongdoers, the belief being that the magnetic iron attracts and draws to it all other particles of nonmagnetic iron. By the same token, if they entered strange premises, the same thing would happen to them and they would be held until the rightful owner appeared.

The author of *Chi-nich-nich*, Dr. John P. Harrington, of the Department of Ethnology of the Smithsonian Institution, found himself against a wall when he wrote something about this "wizard" mineral under its Indian name without a sample from his Indian informant, Mr. Albanes. Dr. Harrington was unable to give a firsthand description and had no definite idea of what the mineral actually was until he requested me to revise his work, a thing I did gladly at the printing office in Santa Ana, Calif.

Later on, when my opinion was asked regarding the book I, of course, drew my own conclusions and withheld any criticism. This I did, not deeming it wise to make any remarks about parts of the work which did not represent the true phases of Indian life, for, as I held the key that would correct those parts, any criticism from me would only have led to my being questioned as to my knowledge of things which I didn't feel at liberty to divulge.

Accordingly, my identification of the "wizard stone" broke up an expensive Smithsonian expedition which was to be made by a group of geologists, with all its branches well-represented, in search of magnetic iron. Now, this is a mineral which has some commercial value in the electrical field, the steel industries and in the manufacture of scientific instruments, although some charlatans, like some of my Indian brothers, claim the mineral possesses supernatural powers and the so-called lodestone is sold today to a good many people as a talisman.

This lodestone is nothing more than common magnetic iron, yellow in color due to the large amount of sulphur in the mineral. The silver-gray magnetic ore which was the Indian's "wizard stone" derives its color from the arsenic it contains. Furthermore, many other spurious contrivances are being sold, not to Indians but to white people of even very superior intelligence and P. T. Barnum fell far short in his estimate of the number of people just waiting to fall into the snare of deception, with his famous remark that "a sucker is born every minute." Every second would be nearer the truth.

By personal investigation which I made I found that some of these deluded victims, men and women both, carry upon their persons lodestones for "luck," whereas others think that by wearing them concealed close to the flesh it gives them a strong magnetic personality and thereby they command the attention they desire from others.

This is not likely to happen. You can magnetize your body, to be sure, but not this way. The right way is the way of the Indians of wisdom, nature's way. If you are getting sluggish, low in vitality and vigor, it simply means that the battery-cells of your body are low and need recharging. Civilization has to some extent deprived us of this great body builder and has isolated us until the flesh of our feet does not make contact with the vast electrical currents of Mother Nature. To receive any benefits from this natural resource you must not wear shoes in the spring or early summer months, and, weather permitting, go to some lonely place where you are sure there won't be any interference by the authorities, remove your clothing and then lie flat on the ground with your muscles well relaxed.

While doing this, dismiss from your mind everything which would cause you worry. Slowly but surely you will find yourself to be a different man if this is repeated yearly. Thus you will have learned to "magnetize" your soul and preserve yourself for the many years before you, as most Indians do.

How to be free from daily annoying trivialities, which only serve to undermine one's health, weaken one's body and render one an easy prey to disease and misery, the Indian understood thoroughly. Instead, he cultivated poise, and reached such perfection that he became the envy of the white race in that respect.

Antivenin for rattlesnake and tarantula bites.

DATURA METELOIDES.
(*Ind. Qui-qui-sa-waal*)

American Jimson Weed. This plant is an inhabitant of the California coastal region and is not particular as to the nature of the soil or its fertility, but thrives anywhere. This tuberous, bushy plant is highly narcotic and when the leaves are properly cured they can be used either in the form of tea or smoked, but withal, very sparingly, since an overdose may very likely cause one to be committed to an insane asylum, as it is a rank poison and its effect may even land one in an undertaker's mortuary. Therefore, my advice is to leave it alone.

My Indian brothers, being unable to give correct information to hard-shelled scientists and writers through a poor knowledge of the English language, were made objects of criticism, and science deliberately declined to acknowledge the medicinal value of this cousin to Datura stramonium which in modern medical practice is of great value at the present time.

Considering, however, the value and uses of *Datura meteloides*, I assert that this decision was very irregular and out of all proportion as to what it was intended. We know that everything can be abused, yet some of our Indian brothers who wished to live in ignorance and superstition had a perfect right to do so. In truth, it wasn't so with all Indians and there were really some great minds amongst them.

When, for instance, the Chinaman wishes to see the beautiful lotus and cherry blossoms of his native land, he does so through the smoke of the opium-pipe.

Then, how about our boasted white civilization which is supposed to be superior to all? Some, of course, like to see Yankee Doodle marching down the street, others draw the hypodermic from its case to stimulate the vision of desire—or sniff cocaine through the nostrils for the same purpose.

How much, I ask, has present-day civilization to offer the Indian? Kind readers and hard-shelled scientists, I pray you, let us be rational and let us go deeper into the field of investigation. And I advise you for your own good to do so even if *Datura meteloides* has failed to make its appearance in the pharmacopoeia with its commercial mark [*Rx] prefixed with the "M.D." As I have said once before, there is very little in the drugstores today for which the Indian cannot find a substitute in the great field of nature. The plants I am writing about in this book represent just one-fourth of the medicinal botany of our Indians of the Pacific Southwest.

In support of my assertion, I only ask that you visit the Historical Southwest Museum in Los Angeles, Calif., and see with your own eyes what a medicine case consisted of for the general practice of medicine in the early days, in those days when most people of the white race thought themselves further advanced in medicine than the Indians. Such was not the case by far, and for a comparison and evidence of my assertion I submit a résumé of the contents used by the first white doctor to practice medicine in the State of California. His name I omit. His soul is at rest!

His medicine case contained a few vials: (1) Oil of Cloves, (2) Spirits of Ammonia, (3) Spirits of Peppermint, (4) Wintergreen, (5) Jamaica Ginger, (6) Castor Oil, (7) Quinine.

And that was all he had to serve as a comparison with what the Indians used, and in this I prefer that you, kind reader, be the judge. I am not exaggerating but speak only in the light of truth. A great deal of misinterpretation of Indian life has been written by white authors who gathered their information from Indians who knew a great deal but were unable to give comprehensive data to the white writer. Others were there who could have done so, but their lips were sealed where the white man was concerned. This accounts for the mistakes which have been made and whereby *Datura meteloides* has failed to gain recognition for its meritorious medicinal value. It is worth knowing that the plant is antivenin and will effect cures of rattlesnake and tarantula bites. What improvement has the hard-shelled scientist made to counteract the deadly poison of the tarantula? None, of course!

Not so long ago a man in New Orleans, La., gave up his life for want of a serum that would counteract the poison of a tarantula which had bitten him. Three doctors, equipped with all the white man's knowledge of medicine and chemistry, strove desperately hard to save the man's life but finally gave up in despair and frankly admitted that to date no serum had been discovered to conquer the poison of the tarantula!

Yet there is one and it behooves the medical profession to get acquainted with *Datura meteloides*.

A practical and effective demonstration was given in the presence of a group of mission friars at San Gabriel, California, in the year 1828. The friars, having heard that *toulache* was used by the Indians to cure rattlesnake bites, lost no time in watching attentively the procedure followed by the Indian in charge—Genio Guana-pia-pa was the medicine man. The patient was seated before them well-covered with a blanket, with only an opening remaining around the neck in order to permit *Datura meteloides* to be poured over his body. All this, of course, was very simple. But what of the process used to extract the poison, when, and what compounds? Yes, there is just one other which belongs in the formula, a formula which has cost me thirty-two years of waiting to finally wrest it from my uncle, Chief Pablo of Pe-we-pe and Guana-pia-pa who gave me access to the ancient historical records of the Indians of California, records which many historians, archaeologists, and practically all men of science would welcome.

The white man has missed the true and better half of the Indian history of the Pacific Southwest. This I shall describe in another book after the translation has been made.

The rattlesnake and tarantula formula I will donate to the good of mankind, free under patent rights under my control to avoid medical exploitation, speculation, and selfish rights of monopoly. I am not in search of mere money. Honor and character are of far more value to me. My happiness comes to me from the good I can do to others, and this to me is almost an obligation to clear and right every wrong done to my people; not with a deadly weapon, but with pen and ink, the Lord willing!

We Indians have not yet learned some of the ways and customs of our white brothers, and, therefore, I could not say that what cannot be done today can be done ten years from now. Our religion and philosophy teach us that the heart is the pendulum of life, each stroke representing timely death which may befall us with one of these strokes any moment, and so don't forget *Datura meteloides*.

Once an incident occurred when *Datura meteloides* was being prepared at a place where the spying eyes of the friars couldn't observe it. Enough plundering had been done already, and the Indians had no desire to divulge their secret. They would only show them what they could do. One of the friars, full of doubt and curiosity, asked one of the Indians who, the day before, had assisted a patient bitten by a rattlesnake to go with him into one of the fields north of the Mission, pretending he wanted information as to certain plants growing there. When there, the friar drew near to a *Datura meteloides*, and at once began questioning the Indian about the plant, whereupon the Indian refused to give information. The friar, seeing he couldn't make him yield, began kicking at the plant, uttering at the same time Latin words, which of course were wholly unintelligible to the Indian. The friar did that only to make the

Indian show resentment, thinking he had cursed a plant which, to the Indian, was sacred.

However, the friar's scheme didn't work; the Indian stood by in silence, until at last the saintly friar lost control of his temper, damned the Indian and his herbs and finally told him that if he'd dig up one of the roots he would chew it, nay, even swallow it, just to show him it was worthless.

The Indian took him at his word, dug up a root and handed it to him. Our friar was game and began chewing it. Accidentally he swallowed some of the bitter juice and in 15 minutes went through terrible convulsions, till death came.

Weather observation, travel and fishing aids, aqueous plants.

ECHINOCACTUS and SEMPERVIVUM (*Ind. Co-pash*)

American Water-Barrel Cactus. An inhabitant of the arid Pacific southwestern deserts. Like all Indians in their boyhood days I was also warned of the great perils threatening human beings in the desert, and in order to meet any emergency which might arise, we received much instruction for the time to come when this was absolutely necessary for our safety. The schooling given us consisted chiefly of weather observation, how to detect the approach of severe desert storms from the very beginning of sunset through part of the night by watching the action of the stars, the night atmosphere and the various changes of the air currents.

All this had to be mastered to such a degree of perfection that I doubt very much if any Bureau of Meteorology could do better. Most of the changes indicated heat waves, electric rainstorms, terrific wind, sand storms, etc. The only instrument the Indian had was his index finger which he made use of by holding it in his mouth for a few seconds, then pulling it out quickly and thrusting it upright toward the sky, his eyes fixed on the North Star. The side of the finger which cooled quickest indicated the direction of the wind.

Now came our water compasses, very primitive in form, but instructive and true to sense of direction as direct leaders to water holes.

Of course the Indians venturing across the arid desert were always careful to take with them the *Sempervivum*—House Leek—and, as an extra measure of precaution, another plant which grows abundantly in the semi-arid regions. These very same things were taught some of the old white pioneers who were friendly and had shown kindness to the Indians, while others perished from lack of knowledge of how to make use of these plants—the *Echinocactus* and the *Sempervivum*, which furnish thirst-quenching juices.

The thorns of the *Echinocactus* were used by the Indians as fishhooks for deep-water fishing so that the modern fishhook is by no means the white man's invention.

Both the *Echinocactus* and *Sempervivum* were also used for the prevention of swelling of the salivary glands as a result of the tongue being dry and inactive.

General medication.

TRICHOSTEMA LANATUM
(*Ind. Zu-bal*)

American Wild Rosemary. This shrub, when in full bloom in the months of May to July, emits a sweet, balsamic fragrance, and is of great medicinal value for many ailments.

The Indians who made use of this plant a great deal had no difficulty in tracing it through its scent to its place of growth, where the flowering stocks were carefully gathered so that the root and crown system suffered no injury. Extra precaution was taken for the next annual blooming season, for most of the plants were of a delicate nature. As far back as I can remember, in the late '90s, the *Trichostema lanatum* grew in abundance along the near coastal ranges, but gradually this very valuable plant became a victim of extermination through brush fires at the hands of careless hunters and the clearing of land by farmers. The Indians, perceiving how rapidly these plants were vanishing, gathered the seeds and carried them further inland, into rough mountain country where they were resown, and there they remained in their last botanical refuge with hundreds of others which are of great medicinal value also.

Furthermore, from such localities Indian hunters would gather the seeds and carry them still further into the mountains for safety and for purposes of propagation. Today it is solely due to the Indian's foresight that *Trichostema lanatum* is found plentifully in the Pe-we-pe mountains, better known today as the San Gorgania mountains. It is also found in Riverside County and the San Jacinto mountains, on southward over high mountaintops in lower California, Mexico, and northward to San Rafael but rarely beyond that point.

Ptomaine poisoning.

ERIOGONUM UMBELLATUM
(*Ind. Hula-cal*)

An inhabitant of the arid California desert, it is a massive, white-flowering shrub remarkable for the long duration of its blooming period which lasts from early June till late September. During this time the desert may be seen covered with a vast mass of white blossoms comparable in its color effect to the winter snowfields in northern latitudes.

To have the opportunity to see the manifold flowering wonders of this great desert, in their sudden magical changes, one must visit it during the period from early February till late in the fall.

The time to see cactus in bloom is in February and March. The latter month ends the flowering season for this particular plant. Again the desert becomes flowerless and gloomy for at least two months and then good Mother Nature with her magic wand once more transforms the desolate desert into a brilliant garden of flowers and shrubs. There are the Yucca Whipplei and the yucca palm, the Joshua palm and the desert lilac, the desert poinsettias, marguerites, desert and scrub pines. All a riot of color, but the *Eriogonum* survives them all into September. What a wonderful array of color greets the eye! As far as it can see it is fairly stunned by these glowing tints and hues, from rich ultramarine to pale yellow—red, lavender, purple, pink, light blue, to the purest cream white. This blooming season ends in July, but alone the *Eriogonum* keeps on blooming till September.

Then there is one short rest period for about one month as if in preparation for a carpet of golden blooms.

Shrubs, bushes, and small trees which have been practically dormant for ten months of the year burst out in golden vestments to greet the approaching winter, and bid a last farewell to Indian Summer.

Such is the aspect of the desert, at that time of the year. And, strangest of all, of all the hundreds of species of trees, bushes, and shrubs, all different from each other, not one bears anything but yellow blossoms.

Let us turn back to where we began, to the especial value of the plant to hunters and vacationists who make their yearly visits to certain mountain and desert regions favored by them. It will certainly do no harm to acquaint yourself with the medicinal value of the *Eriogonum* since you have to depend for your sustenance mostly on canned foods.

Occasions arise when users of such foods are made very ill, even suffer death, through so-called ptomaine poisoning. To obtain the service of a doctor is well-nigh an impossibility and ptomaine poisoning is a fast-working, exceedingly dangerous poison, where delay is fatal.

The plant *Eriogonum* grows nowadays in the coastal regions as well as in the desert where it originally came from, and the first thing to do when attacked by ptomaine poison is to make a strong infusion from the blossoms of the *Eriogonum*, or, if these are not available, use the roots, and take, or give to the poison victim, two brimful cups with a pinch of salt added to each cup.

This remedy counteracts the poison and gives safe and complete relief.

Jewelry and talismans.

ASCLEPIAS SYRIACA
(*Ind. Samat-hap-pac*)

American Milk Weed. White botanists claim this plant to be edible, but to my knowledge there is no botanical record extant which bears out the assertion as to its ever having been used as a food.

The Indians pressed out the milky juices and used the extract obtained in the manufacture of their jewelry, most of their precious stones being made into necklaces, earrings, collars, wrist and upper arm bracelets, all mounted in this milk-juice preparation of the *Asclepias syriaca*.

This kind of jewelry was worn by Indians of a higher civilization as talismans, just as civilized people of today wear similar ornamental articles with more or less superstitious belief. We need only point to the so-called "charms" worn by many to bring luck, ward off disaster or sickness—in fact, for more reasons of a superstitious nature than the Indian ever thought of.

A great deal in the way of royalties are due the Indian, in return for the use of his art-craft, yet a few jewelry items have been overlooked by Mr. Jeweler—items which, peradventure, may bring him great wealth if properly ballyhooed.

Let us take, for instance, the case of the witch-doctor among the Indians with his special ornamental necklace—the symbol of his profession—a necklace denoting supernatural powers, a crystal quartz for a nose-piece. The necklace is made of eagle-, bear- and lion-claws, the poison fangs of rattlesnakes, etc.

It must be understood, however, that the necklace mentioned will not endow the wearer of it with any power whatsoever, unless those various claws are extracted from the living animals which, no doubt, makes the manufacture of such a necklace somewhat hazardous and dangerous.

As far as is known, no patent has been granted yet on the process of getting these necklace ornaments in compliance with the rules of "Charm-Craft." So here is a fine chance for some enterprising, courageous jeweler to strive for renown and riches.

But, Mr. Jeweler, there must be no copying, no pilfering of any sort. The rules must be obeyed in strict honesty. The patent will then be found waiting.

Hunting with poisoned arrows.

KALMIA LATIFOLIA
(*Ind. Po-ha-not*)

American California Mountain Laurel. It may be of interest to sportsmen that this plant, growing in the high mountain ranges, is as greatly relished by the deer as *Hosackia glabra*. Deer in large numbers look for this shrubby tree and this is the key to the white man's puzzle why Indians are such successful hunters. It is simply that the Indian lived the life God intended him to, and, through close association with animals in the wilderness, he became proficient in observing their habits, imitating their calls and thus bringing them within shooting distance of his bow and arrow. Here the *Kalmia* played an important part. The hunter not only used it as a body deodorizer, but also made use of the top part of a deer skull with the skin and antlers left on, never exposing more than these, when lying in ambush.

The *Kalmia* had other virtues, besides, in that it furnished one of the ingredients to compound a poison in which to dip arrowheads. A deer hit with an arrow had a slim chance of getting away wounded. If it happened, it was only for a short distance and then it would drop paralyzed from the effect of the poisoned arrow. The meat was edible and harmless and none was left over.

The grizzly bear was one of the most ferocious members of the bear family in the Pacific Southwest. Although quite common and plentiful, this powerful animal was killed with as much ease as the housewife of today kills a chicken.

When hunting this bear the Indians would select one which had the best coat of hair. Two Indians would work together, beginning with yelling, singing, and dancing around him to such effect that he became very angry and charged one of the hunters. With speed and precision the other one would run up to the bear from behind and shoot his arrow straight into the bear's kidneys, thus in most cases rendering the bear quite helpless; and try as hard as he might to turn on his enemy, his efforts proved futile. Paralyzed in the hind legs due to the arrow being imbedded in the kidneys handicapped him so that he had to vent his rage in a sitting position. The other Indian in front of him, thereupon would shoot an arrow straight into his ear, ending it all in a few seconds. The skin, when cured and tanned, was used for the interior lining of tepees, bed covers, and ground mattresses. However, the time came when bear hunting was abolished by executive orders of our Indian chieftains, and this was caused by the following:

At one time these Indian chiefs, while traversing their territory, were attracted by gunfire. Driven by suspicion and curiosity, they decided to investigate and went to where the shooting was taking place. Great, indeed, was their surprise when they came upon two white hunters battling with a grizzly and it seemed that the bullets of the old-time muzzle-loading gun didn't prove up to expectations. There was very little time to reload, so one of the hunters threw his gun away and fled with the other partner in close pursuit, the wounded bear right on their heels and in full command of the situation.

The sight of this caused the chiefs great mirth and fun, and from that day forth, it was made known to all the tribes that the bear had some human understanding, had no use for the white man and was the protector of the Indians' domain. The act witnessed by the chiefs that day made the bear a regular member and scout of honor of all tribes facing a possible invasion of their virgin country.

In spite of all this, however, the white man resorted to the use of traps and poisoned bait to exterminate the Indians' friend. But, advised by some intelligent instinct this animal had, the bear decided suddenly to leave, and drifted away into Mexico and to northern latitudes.

Even to this day, the bear is considered a great friend by the Indians and when one is killed or dies of natural causes, much reverence and respect is paid him by the older people who, in their minds, are still living in earlier days, now gone by.

Care of the eyes.

SALVIA COLUMBRIAE
(*Ind. Pa-sal*)

This plant belongs to the food division but plays another important rôle, considering what it means to a person to be relieved of the excruciating pain caused by the introduction of a foreign substance into the eye, thus producing a temporary obstruction of the vision. Many Indians, after a hard day's hunting or riding through severe sandstorms, had this experience, and consequently, they never neglected to give their eyes proper care.

When time to retire, the Indian would put at least a couple of seeds of the *Salvia columbriae* under the eyelids, and, with eyes shut tight to keep the seed from dropping out, he would fall asleep. As they swelled, they would move about with every movement of the eyeball and emit a gelatinous substance which gathered up every particle of sand or any other substance present, and, when removed, left the eye clear and free of any possible inflammation. This is a good example of the care the Indian gave his eyes and accounts for his good and strong vision.

An inhabitant of the arid lands, it grows prolifically in places where in earlier days the Indians made their homes, and very often the ancient dwellings which our people used will be found covered with large beds of *Salvia columbriae*, with their beautiful blossoms of purple and lavender.

These colors were to the Indians a mournful reminder of their departed ones, in their lifelong struggle and search for food, so mightily important to sustain life. Regarding the use of the *Columbriae* for this purpose, the method used was very simple.

It was cut and bundled by the male members of the family, brought in and heaped up on a large cleared space of ground, formed into a circle and then trodden down as hard as a cement floor. This was done with water and the bare feet and threshing with long sticks. By thus beating the heap of *Columbriae*, they released the seeds which were then winnowed by being blown before a wind current made with the aid of two baskets. After this, they were carried to the grinding stones to be ground into a fine meal which made excellent porridge—a very popular dish among Indians.

Foods, medicine, tanning and dyes.

QUERCUS VIRGINIANA
(*Ind. Qui-neel*)

American Live Oak. This evergreen tree of the western mountain ranges is the most imposing of all the species of the oak family. It grows to an immense size and attains a great height. Some of these giants of the forest cover and shade an area large enough to afford protection to as many as three hundred adult persons.

A great deal has been written in song and poetry in praise of the stately oak but the Indians found out by experience that the acorns it bore were far more nutritious than poetry, and before long the noble tree was adopted as a regular member of the tribes—a bountiful provider of food.

Even so, the oak was by them much honored in war and love songs, for the many good things it furnished them besides food. The fallen leaves made warm mattress-bedding while the bark played a part in medicine and also in the tanning and dyeing of buckskin in various fast colors by blending with the bark of other oaks and roots. Let it be understood that these dyes thus produced were of a firm, non-fading nature and also excellent preservers of buckskin.

The colors produced were very beautiful and ranged from pure white to yellow, red, light and dark brown, light pink, gray and black.

Regarding the acorns, special care was given to the harvested crop and the process was simple. The acorns were put into fine, hand-woven net bags and tied with a rawhide rope to a tree close to the river bank whereupon the bags were placed in the stream. The running water would cause the acorn shell to swell and split open, thus releasing into the water most of the tannic acid which the acorns contained. After being left in the water for a week or so, they were taken out, the hulls removed and spread out to dry. Afterwards they were ground into a fine meal, sun-dried again, and then put away for winter use.

The porridge made of it jells like custard and, when well cooked, has the color of chocolate pie. It can be cut into squares and served with deer meat or eaten as a dessert with cream and sugar. Besides being very delicious and nourishing it is also a great flesh builder.

As a warning, let it be said, never to eat any acorns picked fresh from the tree, because of the tannic acid they contain; in that state they may cause severe constriction of the bowels and the glands of the throat.

Bleeding navel.

TYPHA LATIFOLIA and QUERCUS AGRIFOLIA
(*Ind. Co-o-tem*) (*Ind. Qui-neel*)

American Cat-tail is an aquatic grass inhabiting shallow, stagnant lakes and swamps and is very common on the Pacific coast of California. *Tule* is perhaps the name by which the plant is best known, although the other is also very common. This valuable grass has failed to find a place among the scientists of the world, as *Tule* is a purely Indian name, and is far from being identical with those so far being classified by botanical science. But *Tule* is medicinal and has healing properties which were made use of by the Indians to heal bleeding navels. Nothing could be better.

The blades of the grass were gathered and burned to the consistency of charcoal, then finely powdered and sprinkled on the bleeding parts.

When this couldn't be obtained, the Indians further inland had recourse to the apples growing on the Scrub-Oak or *Quercus agrifolia*, and these were, of course, dried and powdered, and medicated with balsam oil. The salve proved to be very effective in healing the afflicted parts. In short, the results were first-class and saved the lives of many little Indian babies.

Indian food.

PROSOPIS JULIFLORA
(*Ind. Pe-che-te*)

Mesquite Bean. An inhabitant of the southwestern deserts, it ranges as far as the northwestern and southwestern central parts of Mexico. A native of southeastern California, Arizona, New Mexico, and Texas, the *Juliflora* was perhaps one of the trees which provided the greater part of food for the natives.

Its contents were very rich in protein and even wild animals relished it greatly. To obtain the yearly supply, the Indians made a regular pilgrimage early in the season to localities where the *Juliflora* grew in abundance, and stood guard over the trees for many weeks until the bean pods were fully matured. Then they were harvested and ground in rock mortars to the fineness of flour, such as is used in the baking of cakes, tarts, etc.

It could also be mixed to the consistency of porridge, either with hot or cold water and taken with sun-dried venison. It formed a very nourishing diet. Sugar was never added to it.

The bean pods of the *Juliflora* are extremely sweet, and may be eaten right off the tree if dry enough. In any other condition they are unpalatable.

PTERIS AQUILINA
(*Ind. Wel-met*)

American Bracken Fern. This graceful and stately fern of great beauty of leaf design inhabits the high mountain ranges where there are well-shaded forest lands rich in mulch. This fern is well-known to every Indian for the sad historical part it played in the life of our fair and beloved sister Ramona, the daughter of Ca-we and wife of Alessandro, the immortal Indian who suffered death without a moment's warning at the hands of a brute and coward.

The authoress of *Ramona*, Helen Hunt Jackson, mentions in her book what good use of this fern Alessandro made in preparing Ramona's bed at the time of their elopement and tells of the hardships both young lovers underwent.

The young sprouting shoots of the *Pteris aquilina* fern mean as much to the Indians as asparagus does to white people, as it contains much oil which is extremely rich in flavor when the shoots are properly cut and cooked.

There are ferns in song, ferns in poetry, ferns where wedding bells ring, ferns on the altars of churches and ferns in God's acre. Also in gardens, but nature's garden is where the Indian wants them!

Food and bleaching.

YUCCA WHIPPLEI
(*Ind. Yu-ca*)

American Spanish Bayonet. The name yucca is the true native Indian name of this exquisite plant, but even Mr. Whipple, the botanist, failed, like many others, to properly describe the beauty of the yucca.

During the months of May and June when the plant is in full bloom it is nothing strange, when venturing into the desert mountains, to find oneself in a veritable forest of countless thousands of yuccas. With its erect stalk, attaining a height of from four to twelve feet, heavily and massively crowned with creamy white blossoms, the yucca closely resembles a gigantic hyacinth of the California desert and mountains, and its delicious fragrance outrivals many of the costliest perfumes.

The use of the yucca was of much importance, some of the stalks were cut just at the time the plant was in full bloom, the flowers are edible, the stalk rich in sugar which produces a fine quality of syrup, obtained by first roasting the stalks in underground pits.

Other stalks were allowed to mature, their pods yielding the finest material for bleaching buckskin fiber a pure white. Also used very much in the art of basketry, etc.

Rheumatism.

URTICA HOLOSERICEA
(*Ind. Panga-tum*)

American Stinging Nettle. An inhabitant of the swamps and river beds. This plant was used in most cases of inflammatory rheumatism of the most peculiar kind known to mankind, particularly when the lower limbs were affected to such an extent that they became numb, cold and useless. The cure was very simple if your limbs were in a bad state, but rather unpleasant if in a sound condition.

The nettle was cut and brought to the Indian patient's bedside, where the leaves were rubbed on all his ailing parts. This was repeated for several days until warmth in the affected parts and a proper circulation of the blood was attained. When the patient was able to get up and walk, a second treatment of a different nature was administered, the so-called Rock Steam-Bath of a herb compound made up of the following three plants:

PHLOX SUBULATA
(*Ind. E-wa-yack*)

American Moss Pink. An inhabitant of the Mojave Desert.

PINUS MONTICOLA
(*Ind. Wa-ta*)

American Scrub Pine. An inhabitant of the northern slopes of our California Mother coast range, and in a few localities on the desert floor.

ADIANTUM CAPILLUS-VENERIS
(*Ind. Ta-wal*)

American Southern Maidenhair Fern. Inhabits the high coastal ranges, but further north it will be found on the lower coastal ranges.

Menstrual period.

LIPPIA LANCEOLATA
(*Ind. Te-eel-p-yack*)

Lemon Verbena. Spanish *Cedron*. This shrubby tree has become nearly extinct and but few specimens are found now and then. The infusion made from its leaves and blossoms is very aromatic, somewhat like peppermint.

CRYSANTHEMUM PARTHENIUM
(*Ind. Che-ke-wat*)

American Feverfew, Spanish *Artemisa*. This plant was used for the same medicinal purpose as the one mentioned above.

Diseased throat glands, scrofula.

NICOTIANA GLAUCA
(*Ind. Tee-baat*)

American Tobacco Tree. This tree, very common along the Pacific coast, grows from Santa Barbara southward to the end of Lower California, and the Mexican peninsula. The tree grows in terraced gorges and ravines and is rarely to be found anywhere else.

The leaves were steamed and applied externally as a poultice over the swollen parts of the throat caused by inflammation of the throat glands, and also for scrofula. While the latter malady didn't exist among the Indians, yet they treated and cured some of the whites who had it, with *Nicotiana glauca*.

It was also steamed into the body of those suffering from rheumatism and proved there also its value to many human beings.

I have once before spoken of other plants useful for the same purpose, but as this plant has something to recommend it for the last-named ailment, it is appropriate to mention it again in connection with scrofula and inflammation of the throat glands.

To the plant serving all these cases equally well, we must give credit where credit is due, even at the cost of repetition, in order to give the reader a fair understanding of the various diseases a plant may be good for.

Surely a wonderful provision made by nature!

Fishing.

CROTON SETIGERUS
(*Ind. Tu-tal*)

American Dove Weed. The beautiful dwarf plant is very common throughout the coastal region and far into the inland valleys. It appears about July in most barley fields after the harvest. It is truly a paradise for wild turtledoves, and the hunter who goes into a place where the *Croton setigerus* grows may be sure of bagging a good number of doves in a short time.

The Indians gathered the plant for use in their fishing operations, and some of it was stored away for winter use. The weed has a strongly intoxicating effect on fish.

A place was selected along the stream bed in a rather shallow spot and dammed across.

After this, a regular mat, formed of *Setigerus*, was laid on the surface of the water, while a large number of Indians went upstream to herd the schools of fish downstream and into the trap. Quite a simple procedure, as the herding was done by merely beating the water ahead of them. A barricade built of brushwood behind them prevented the fish from going upstream. The water in the pond having become impregnated with the *Setigerus* affected the fish so that they soon floated helplessly on the surface of the water where the Indians just picked them out by hand. When a sufficient supply had been taken, the *Croton setigerus* was removed and piled up on the bank of the stream to dry and be used again. The dam and barricade were also done away with and the uncaught fish were allowed to get into fresh water to recuperate.

Tonic for loss of appetite.

MONTIA PERFOLIATA
(*Ind. Lah-chu-meek*)

American Miner's Lettuce. This plant inhabits the coastal regions where it thrives only in deep, decomposed beds of oak-tree mulch at suitable points in the shady woodlands, where the circulation of water is present under a deposit of mulch.

The juice of the plant is an excellent appetite-restorer.

ALLIUM BISCEPTRUM
(*Ind. Ye-sil-ta-usa*)

American Wild Onion. It is an inhabitant of the lower mid-coast ranges, and the extract obtained from it is compounded with the powdered berries of *Rhus trilobata*.

RHUS TRILOBATA
(*Ind. Sa-lat*)

American Squaw-weed. An inhabitant of Southern California's higher ranges, it makes an excellent restorative for an inactive stomach which refuses food. The Indians also obtained the fiber from the vines of the shrub by stripping it off with the thumbnail and using it for basket making.

For poisonous insect-bites.

ALLIUM CANADENSE or ALLIUM VINEALE
(*Ind. Ye-sil-we-na*)

American Wild Field Garlic. A plant held in great esteem by the Indians, protecting them, when hunting or exploring, from poisonous snakes, lizards, scorpions, tarantulas and insects during the summer season.

It was the custom of the Indians then to discard their buckskin clothes and roam around with as little covering as possible until the fall of the year, when they donned their heavier clothing again for the approaching cold weather. Now, it is well-known how disagreeable the odor of garlic is to most human beings, but they don't know that it is likewise so to reptiles and insects. The Indians, however, knew this, although they never ate it. They used it only as medicine when needed, but its greatest usefulness was to guard against being bitten by poisonous vermin.

The Indians ground the wild garlic into a pulp and then rubbed it well over their legs up to the thighs, making extra sure that the skin was thoroughly saturated with the garlic juice and thus protected. The Indian would enter any locality to do his hunting, even if it was infested with thousands of rattlesnakes, without the slightest fear or worry. The reason is very simple. Whenever the snake or insect comes within smelling distance of the garlic, it is so much affected by it as to become well-nigh asphyxiated and is rendered helpless.

The white man, in order to follow fashion, wears leggings, but I am sure that he could use the formula I have given, very much to his advantage. I give this formula freely to mankind, a formula which has remained a secret for over a century and it will mean the saving of many lives if used as described above.

Antidote.

BERTHOLLETIA
(*Ind. Pacah-quit*)

American Arrow-wood. It is an inhabitant of the California River border lands within the Pacific coastal belt, and is occasionally also found on the southern border of the western desert lying in the northern part of the Pacific coast.

There has been much discussion in the past, and many arguments, many flatly declaring that the arrowwood was used by the Indians for making bows and arrow stocks.

Being an Indian, that and nothing else, let me explain the matter clearly as to this particular controversy. The young shoots of the *Bertholletia* were selected from the parent stock, well-seasoned and then used for arrow stocks on which small arrow points were fitted for the young Indian children to practice and hunt with. It was never used for bows, however. For the making of fire through friction, it was very useful and yet, this alone would not give an adequate account of the value of arrowwood shrub. This is left to the decoction made from it, to counteract the poison in wounds inflicted by arrowheads in battle engagements, and therein lies its principal claim to the consideration shown it by the Indians.

Sedative.

PHYTOLACCA DECANDRA
(*Ind. Che-ne-va-ica-cal*)

American Ink Berry. This shrub, a common inhabitant of California's coastal regions, has been placed by the white writer in the division of poisonous plants, and we agree with him. So the only credit given the plant is chiefly for the remarkable beauty it displays with its starlike flowers and racemes of dark-blue berries. Yet it has been condemned under the label of poison, and much is being done toward its destruction wherever found. However, it is a fruitless task, and may only become a near-success when the Indians and the birds shall be known as two signs of life vanished from the face of the earth. For these two are responsible for the preservation and propagation of the shrub.

Morphine, opium, and cocaine are by far deadlier poisons than *Phytolacca*—why, then, do doctors prescribe them to soothe and ease pain, etc.? The root of the plant has some medicinal qualities to ease severe neuralgic pains, and is deemed very efficient and important in Indian medical formulas. For making dyes and inks the berries are excellent, whereas the leaves are most useful in the treatment of skin diseases, and to eradicate and clean the epidermis of pimples and blackheads.

Therefore, help to conserve and not destroy this really valuable plant.

Diseases of the liver.

RORIPPA NASTURTIUM OFFICINALE
(*Ind. Pang-sa-mat*)

American Water Cress. It is an inhabitant of the coastal regions, swamps and rivers. This aquatic plant is more deserving of attention than has been given it, and is fully worthy of the name it bears, *Officinale*, which means all that the word implies. The Indians, having discovered the medicinal qualities of this plant, immediately gave it a place in their medical and food division and, up to the present year of our Lord, the plant has been used in the treatment of disorders of the liver—cases such as torpid liver, cirrhosis of the liver and as a dissolvent of gallstones, etc.

When these diseases are curable, the diet is simple—with no restrictions and no red tape to plague the patient. The first meal taken in the morning must consist of *Nasturtium officinale*, salted very sparingly, and of this the patient should eat as much as possible and do without further food until noon, when he may eat whatever he likes. This method must be repeated every morning. Care must be taken not to use liquor if one wishes to insure quick recovery.

When the liver is ulcerated it takes at least two months to heal properly, but all other cases are of short duration.

Reducing teas.

LEPIDIUM EPETALUM
(*Ind. Chesa-mok-ka-mok*)

American Pepper Grass.

SALINIA
(*Ind. Cheena-wah*)

American Salt Grass.

PANICUM CAPILLARE
(*Ind. Ne-wa-cha-mo*)

American Witch Grass. The first two are fond of rich, agricultural soils, whereas the latter prefers alkaline lands. All three have been declared noxious weeds and are listed as such by the Department of Agriculture, although the Indians found some use for these grasses.

There were times when some of our men and women became over-fat; in fact, so fat that they had great difficulty in traveling, the exertion making them complain of heart trouble which in reality was nothing but a discomfort due to short respiration caused by excessive fatness. Accordingly, something had to be done. A search was made, and experiments with good results finally obtained. These grasses compounded with the bark of sassafras, wall-wort and others (also named for extermination, just like the three above-named plants) are excellent for reducing purposes.

The chief trouble in our schools where botany is taught seems to be that too much attention is given to the *appearance* of plants, instead of to their medicinal value and other useful properties.

Birth control.

IVA AXILLARIS
(*Ind. Na-wish-mal*)

American Poverty-Weed. This hardy plant predominates on most of the salty marshes and lake shores. It is hardly worth destroying as it mostly grows in soils totally unfit for agriculture, or anything else, for that matter.

Let me mention, however, that there is quite a history connected with the earliest beginning of the Indian's life in connection with this plant. No doubt it will be of interest to the readers of this book to learn that the plant played an important part in what is today assumed to be a modern institution—birth-control.

The Indians knew and practiced it from the earliest times, but only in cases when women proved themselves incapable, even when at their best, to give birth to healthy children.

In such cases they were compelled to make use of this plant as a preventative and this should explain the Indian's wonderful stamina, his sturdiness and perfect physique. Moreover, the great chiefs prohibited the raising of deformed children, as ordinarily they considered this a great sin.

In later years the secret was let out by some Indian women, and thus it found its way among the Spanish and American settlers, when many cases of abortion were due to the use of this herb—a universal practice of modern civilization with its accompanying evils of genocide and other evils of a criminal nature.

Kidney diseases.

CROTON CORYMBOSUS
(*Ind. O-chot-pa-wish*)

Spurge. Its habitat is in the southern Mojave sand dunes. This beautiful shrub, like many of the other desert plants, seems to select the worst of soils to grow in, and is often to be found in crevices of mineralized dykes of crystalline rocks. The infusion made from the plant cured kidney infections.

EPHEDRA
(*Ind. Tut-tut*)

The Tea of the Indian is found in the swamplands of the coastal regions. The infusion made from the leaves and blossoms was taken internally for pleurisy of the kidneys.

An infusion made separately from the roots was also used internally to relieve severe cases of gonorrhea and painful bloating of the stomach. This remedy is very effective and highly esteemed by the Indians as one of the royal plants for the cure of these dangerous ailments, which take the lives of so many of the white race.

APIUM
(*Ind. Se-ma-mek*)

American Parsley. Its habitat is the swamps and coastal regions. The infusion made from this plant was taken regularly and in preference to water or any other beverage for chronic diseases of the kidneys.

The tea is very rich in flavor and pleasant to the taste. The patient should partake of as much as one half gallon per day and also eat an equal amount of it. The plant having been domesticated it is no trouble to get it anywhere. Even butcher shops and vegetable dealers sell it.

XANTHIUM CANADENSE
(*Ind. Cho-co-late*)

American Cocklebur. It grows everywhere in California, being found in every swamp and pasture land—a veritable nuisance to the cattle raiser.

From the medical standpoint, however, the plant is very valuable to the members of both sexes who are suffering from diseased kidneys complicated with gonorrhea, diseases which, when allowed to take their own course, will in due time develop into tuberculosis, rheumatism, and finally total paralysis of both the upper and lower limbs, as has happened in such cases.

The introduction of these maladies occurred with the advent of the white race into our territory and this caused the Indians to go into further botanical research to find the proper plants to combat and conquer these dreadful diseases. I introduce the world in general to two other sister plants, and also three belonging to a different group.

CENTAUREA MELITENSIS
(*Ind. Se-sa-naa*)

American Star Thistle.

XANTHIUM SPINOSUM
(*Ind. O-yu-mo-val*)

American Spiny Cocklebur.

MALVACEA RUBRA
(*Ind. E-ya-wa-manka*)

American Creeping Rock Mallow. Spanish *Yerba Mora Real.*

Venereal diseases.

CERCOCARPUS BETULAEFOLIUS
(*Ind. Man-geet*)

American Mahogany Shrub. Its habitat is in the California hills and mountains, and it is quite common. The bark and roots were made into an infusion and taken by the Indians for venereal diseases or gonorrhea gleet.

CENTUNCULUS
(*Ind. Pepe-nel*)

American Pimpernel. Its habitat is on the northern slopes of the highest mountain peaks of California, at an elevation of from eight to ten thousand feet above sea level. This wonder plant is made into a tea and taken in acute cases of gonorrhea, where the bladder and urinal tract fail to function.

EDIBLE FRUITS OF SHRUBS

The plants listed here are common in our California mountains:

Arctostaphylos	Manzanita Berry
Sambucus pubens	Elderberry
Ribes glutinosum	Wild Currant
Ribes amarum	Gooseberry
Prunus serotina	Wild Black Cherry
Prunus ilicifolia	Hollyleaf Cherry
Heteromeles arbutifolia	California Holly Berry
Vitis vulpina	Wild Grape
Rubus villosus	Wild Raspberry
Rhus trilobata	Squaw Bush Berry
Rhus integrifolia	Lemonade Berry

These berries should be eaten sparingly, as the acidity contained in them is much stronger than that of citric acid. Their chief use is to quench the thirst, where water is scarce in the mountains, either when hunting or hiking, or engaged in fighting forest fires. For this purpose the berries above will be found excellent and a veritable boon. Everyone traveling in desert or mountains should make himself familiar with the plants and fruits growing therein, as this knowledge not only permits him to guard against possible discomfort or hardship, but has also been the means of saving life. The Indians knew that better than anyone else.

No doubt, the following literary effort in the English language by Chief Pablo will set the risibilities of my readers to working. Als, the Chief, never had the benefit of a school education, and English wasn't easy for him to acquire. However, he was game, and in 1908, when he was appointed Chief of the Indian Reservation, he bravely set to work and wrote this article.

He was sixty-four years old then. Nothing would do but he must have a typewriter. Right manfully he tackled it, but when he had finished he heaved a tremendous sigh and declared he'd rather go on the warpath than pound a typewriter again.

But he surely deserves great credit and his record as Chief of the Indian Police was a brilliant one.

THE LEGEND OF CONSOLE MINERAL SPRINGS NEAR HOMUBA CANYON

The canyon has been known as Homuba among the Indians for many years. And on that canyon there are three mineral springs. They are located near Loma Linda. It is southeast from Loma Linda, way up in the canyon, a distance of three miles.

Professor J. Console, an Indian friend, is the owner of the mineral springs nowadays. In the early days the Indians called the springs *Phal-poole*, *Phal-quapekalet*, *Hickescah-heppasca*, which means Witch Springs, Life Springs, Sisters and Brother Springs.

Those three springs were discovered in this way. There were Indian settlements all over that country, near the springs and around the springs. One day three Indian children, two sisters and one brother, went up the canyon and disappeared in those springs. The father and mother and other relatives of the missing children followed the tracks of the children until they came to the springs. After having tried everything to find them, the father and mother and the relatives in their sorrow went to the witch-doctors to see if they could help them find the children. Then one witch-doctor said:

"Come with me and I will show you where your children are and how they disappeared in those springs. You may not see them but you will hear them, and you will have to be satisfied."

So the children's family went there and the witch-doctor stopped at the center spring and said:

"Listen to the Great Father who is above us, the Creator of the world. He has taken your boy and put him in this spring, so that this spring will bring health to you and to others."

Then the witch-doctor walked up to the spring and spoke:

"Brother, your father and mother and all the relatives are here, and they would like to hear from you."

Then a voice arose from the spring and said:

"I am here with my two sisters. We were placed here by our Lord, the Creator of the world. He has given me the power to bring new life to those who are sick. You may come and visit me and my sisters whenever you wish. My elder sister is in the spring on the east side of me, and my younger sister in the spring on the west side of me. But we are all three in this one place, and if you will live together and honor the great Lord, when you are sick if you will use these life springs, we will help you get back your health. These springs shall be known as the 'Two Sisters and Brother Life Springs.'"

And all the people listened. Therefore the Indians went up there and held a great ceremony, and from then on they used the springs for medical purposes.

Then the Catholic missionaries came to this country and established the missions. They took the Indian children by force and made them Catholics. And these Christians also went up to the springs and used them for many years.

Later on the United States Government came to this country and took these lands and gave the Indians reservations for their use. And the Indians had to leave the springs, which originally belonged to them.

When the mission was first built at San Gabriel the priest asked an Indian:

"Why do you Indians take your children, when they are sick, to those springs, instead of taking them to a doctor?"

And the Indian answered:

"Father, the springs at Homuba Canyon can cure any sickness. That is why we take our children there when they are sick, and they are healed. Our ancestors used those springs and became healed."

Then the priest went up to the springs to examine the water, and he took some of the water and made the Indian carry it to the chapel, and he blessed the water, and held Mass with it, and used it to cure the sick. And, finally, the priest moved the mission from San Gabriel to San Bernardino. Old San Bernardino is now known as Redlands. The mission was established there. It is about three miles from the springs. And from there the priest used to send the Indians to bring the waters to the mission, using it as medicine. And he cured many sick Indians.

Now there were two Indian villages nearby, and they fought over the possession of those springs. They went on the warpath over the Two Sisters and Brother Life Springs. So the mission went away and settled elsewhere, and the priest also went away.

Then our white neighbors came, as I said, and drove the Indians from our sacred springs. That is why the Indians are dying out in Southern California, because we must live on worthless lands far away from those springs.

Our white neighbors may think we Indians have no religion, but that is not so. We do believe in God who is the Creator of the world, and of the firmament, of Indians as well as white people.... Therefore we are brothers in God, as we are created by God.

I often hear white people say they are Americans in America, and we are Indians. I say we are the native sons of America. We are good to our country and to our white neighbors, and do not trouble them. When the missions first came to this country the Indians were numerous and the country well inhabited by the Indians. Then the Indians did not know that the country was going to be filled with intoxicating liquor. If they had known that, they would never have allowed the missionaries to establish any missions in this country. For a great number of Indians died of intoxicating liquor.

However, the United States Government made a law prohibiting the sale of intoxicating liquor to the American Indians. But by that time it was too late. The American Indians are nearly all gone. But maybe a few will be saved.

It was in 1908 that special officers suppressed the liquor traffic among the mission Indians in Southern California. The chief of these special officers came to me and asked:

"Why is it that you are always fighting the whites?"

"Because they are all liars, thieves, and whisky peddlers," I answered.

He looked at me and said:

"Am I a liar and a whisky peddler?"

"No," I answered. "You do not look like one. I think you are on the square."

So he said to me:

"I want you to work with me on the same job."

"What job do you mean?" said I.

"To suppress the liquor traffic among the mission Indians," he said.

So I was deputized as Special Officer since then, and I became Chief of Police in the Indian Service for nine Indian reservations under the United States Government, to protect the Indians, to make transactions for the Indians, and to help them become sober, improve their morals, and become civilized. In 1847 if the United States Government had sent us a man like Mr. C. T. Coggeshall, who is the superintendent of the nine Indian reservations, the Indians would never have lost the Two Sisters and Brother Life Springs. Mr. Coggeshall is a man with large experience and he has done a lot of good for the Indians under his jurisdiction.

However, I am glad that Mr. John Console owns the springs, because he is a friend of the Indians. He helps the Indians with those springs. The springs cure light sickness, but for serious sickness we have to use herbs.

CHIEF WILLIAM ALS PABLO

THE BOTANICAL LORE OF THE CALIFORNIA INDIANS

by

John Bruno Romero (Ha-Ha-St of Tawee)

Rare Indian lore collected and interpreted by a full-blooded Chu-Mash Indian, who grew up among members of the Cahuilla tribe, is revealed in this unique book. Written by a man who is anxious to share his ancestral knowledge of the treasures in the Great Field of Nature, this volume describes 120 medicinal herbs and gives recipes for their preparation, their uses, their English and Latin names, and where they may be found.

The collection presented here was hand-picked from 500 specimens gathered by the author on a plant-hunting expedition on the Pacific Coast and in Arizona. Only twenty-eight, it is said, are known to modern medical science.

For more than one hundred years, the Indians have kept to themselves their profound knowledge of medicinal herbs and their application. Meanwhile, if the Indian, with his intelligent and extraordinary attachment to nature, had not preserved and replanted a large number of these herbs, many of them would now be extinct.

A close collaborator of the historical department of the Santa Ana Museum in his native California, the author is known as a botanist of such high order that some years back the British Museum sought his assistance in assembling a remarkable collection of Pacific Coast specimens of medicinal herbs and Indian artifacts.

Mr. Romero, whose Indian name is Ha-Ha-St of Tawee, presents his material in highly entertaining manner, and his remarks, some of them *sotto voce*, are extremely apropos. Adding to the color of the book is a wonderful legend written by the author's father, Chief Als Pablo, Chief of Police of the Indian reservations in the Southwest at the turn of the century.

The book is dedicated to the memory of the author's uncle, Chief William Pablo of Guana-pia-pa, medical herbalist and medicine man. It is a unique treasury of authentic Americana, fortunately preserved for our time.

VANTAGE PRESS, INC.
120 W. 31st St., New York 1

JOHN BRUNO ROMERO

John Bruno Romero is a descendant of the Chu-Mash, once the largest and most powerful of Indian tribes, whose domain included all the islands along the California Coast, and, on the mainland, from the San Fernando Mission northwest to San Francisco and north-northwest to the High Sierras.

Mr. Romero was born in Santa Barbara, where he studied Spanish and Latin at the Franciscan School Mission, and attended the Sherman Institute, where he was a student of English and scientific subjects. He was later graduated from the Detroit Veterinarian College.

At one time, the Cahuilla Indians controlled the lands of California southward to the end of what is now the Mexican peninsula. When the Chu-Mash tribe, in its later years, had dwindled in numbers, Mr. Romero joined the Cahuilla tribe "to help fight the United States Government for our land treaty rights." Today this tribe is the second oldest in California and the strongest in membership.

The author's interests are wide. He is a director of Indian Affairs for seven Southern California counties, and while his principal hobby is medicinal botany, he is also a collector of minerals, stamps, books, and fossils, and dabbles in taxidermy. Fond of children, he has adopted and reared ten orphans. At one time or another, he has worked as a surveyor, explorer, geologist, and antho-botanist, and his home is a veritable treasure trove of interesting archaeological, geological, and botanical specimens which he has collected in the mountains and deserts of Southern California and Arizona—in sections where the white man has seldom traveled.

In 1933 he discovered, in the Trabuco Hills, in Orange County, near Los Angeles, the skeleton of a mastodon, one of the few uncovered in Southern California. Paleontologists at the Los Angeles Museum made varying estimates of the age of the bones, ranging from ten thousand to a million years.

In 1937, when the author was a junior geologist at the Santa Ana Museum, he deciphered the hieroglyphics inscribed on some rocks found on the Indian prayer grounds at the peak of a volcanic vent in La Piomosa range in Arizona, which have been instrumental in providing historical information about the life of the early Indians in the Southwest. The inscriptions revealed the location of food and water in the surrounding country and primitive conceptions of the supernatural.

Transcriber's Notes

- Retained copyright information from the printed edition: this eBook is public-domain in the country of publication.
- Silently corrected a few palpable typos.
- In the text versions only, text in italics is delimited by _underscores_.

www.ingramcontent.com/pod-product-compliance
Lightning Source LLC
Chambersburg PA
CBHW070859280326
41934CB00008B/1506
* 9 7 8 1 6 3 6 5 2 2 8 4 5 *